MICHAEL J. BENTON

Dinosaurs
Rediscovered

The Scientific Revolution
in Paleontology

Dedicated to my wife, Mary, and
children, Philippa and Donald, for putting
up with me.

Dinosaurs Rediscovered © 2019
Thames & Hudson Ltd, London

Text © 2019 Michael J. Benton

First published in 2019 in the United
States of America by Thames & Hudson
Inc., 500 Fifth Avenue, New York,
New York 10110

www.thamesandhudsonusa.com

Library of Congress Control
Number 2018956110

ISBN 978-0-500-05200-6

Printed in China by Shanghai Offset
Printing Products Limited

Dinosaurs
Rediscovered

Contents

Geological Timeline

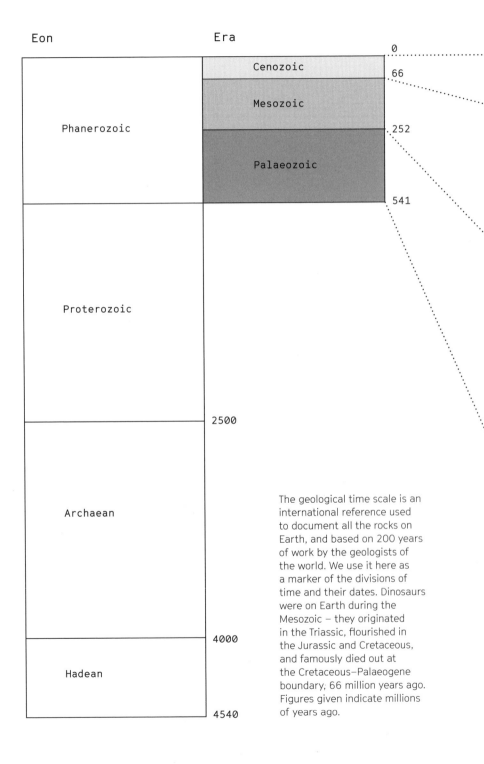

Eon	Era	
		0
Phanerozoic	Cenozoic	66
	Mesozoic	
		252
	Palaeozoic	
		541
Proterozoic		
		2500
Archaean		
		4000
Hadean		
		4540

The geological time scale is an international reference used to document all the rocks on Earth, and based on 200 years of work by the geologists of the world. We use it here as a marker of the divisions of time and their dates. Dinosaurs were on Earth during the Mesozoic – they originated in the Triassic, flourished in the Jurassic and Cretaceous, and famously died out at the Cretaceous–Palaeogene boundary, 66 million years ago. Figures given indicate millions of years ago.

Era	Period	
		0
Cenozoic	Neogene	23
	Palaeogene	
		66
Mesozoic	Cretaceous	
		145
	Jurassic	
		201
	Triassic	
		252
Palaeozoic	Permian	
		299
	Carboniferous	
		359
	Devonian	
		419
	Silurian	444
	Ordovician	
		485
	Cambrian	
		541

How Scientific Discoveries are Made

Discovery

I can remember the day – 27 November 2008 – when Paddy Orr came through from the scanning electron microscope (SEM) lab in Bristol, and said 'We've found these regular organelles in the feathers. What do you think they are?' I went through, and he, I, and Stuart Kearns, who runs the facility, checked over the tiny chippings from the feathered dinosaurs from China. There they were on the screen – rows of slightly distorted spheres deep in the feather tissue. As Stuart rolled the control ball, the field of view changed and wherever we looked there they were...

Melanosomes.

In a 125-million-year-old fossil feather.

Spherical melanosomes in a fossilized feather of the dinosaur *Sinosauropteryx*.

Melanosomes are the tiny hollows inside hairs or feathers that contain melanin. Melanin is a pigment that gives the black, brown, grey, and ginger colours to hairs and feathers. We were the first ever – or at least the first on record – to have seen evidence of melanosomes in dinosaurs. If we had got it right, this was evidence of the original colour of their feathers. We could say that for the first time we had discovered for sure the colour of a dinosaur.

We were torn by emotions at this point. Our first desire was to rush out and tell the world – call the press and shout from the rooftops! On the other hand, as scientists, we are trained to be careful, and we wouldn't want to look foolish by making such a wild claim if the evidence wasn't there. There's also a whole process behind publishing science, the so-called peer-review process, which ensures you present all your evidence, in detail, and sufficient to pass scrutiny of two or three independent colleagues. Only after publication in a scientific journal do you release your discovery to the mainstream media.

So, we went out for a beer, and planned to look at more specimens and make more measurements. This was a hugely controversial observation back in 2008. The microscopic structures could be melanosomes, as we thought, but the critics would shred us if we couldn't show multiple observations, and rule out all possible alternative interpretations.

Over the past thirty years, opinion has moved back and forth – these tiny structures in the feather tissue were interpreted as bacteria, or artefacts, or melanosomes…Sometimes they were like tiny balls – as here – and sometimes like tiny sausages in shape. At one micron or half a micron across (a micron, or micrometre, is one-millionth of a metre or one-thousandth of a millimetre), we were working close to the limits of the magnification capability of the SEM. Was there any way they might be inorganic artefacts, perhaps some mineral crystals that had entered the feather during its fossilization?

Earlier that year, Jakob Vinther, Danish by birth but at that time a doctoral student at Yale University, had published an important paper that showed how the micro-balls and micro-sausages in fossil bird feathers occurred only in dark-coloured areas in the fossil – they were melanosomes, not bacteria. He argued very convincingly that if they were bacteria that had invaded the feather to feed on minerals in the decaying specimen, they should be distributed equally all over the surface, on both dark- and pale-coloured stripes.

We accepted his view and immediately applied this brilliant insight to fossil specimens we had been working on with our colleague

Dr Fucheng Zhang from the Institute of Vertebrate Paleontology and Paleoanthropology in Beijing. Fucheng had been a postdoctoral researcher in Bristol in 2005; he had brought over examples of fossil feathers from dinosaurs and birds, and we had been studying them.

The feather chippings came from *Sinosauropteryx*, a slender 1-metre-long (3-foot) dinosaur with a long tail and short arms – not a flyer. But the *Sinosauropteryx* fossils preserved beautiful examples of whisker-like feathers along the back and as tufts down the tail. Melanosomes, we knew, were the hollows in the keratin protein of a feather into which the pigment melanin is inserted as the feather grows. Ball-shaped melanosomes in our samples showed *Sinosauropteryx* was ginger – it had a neat ginger and white striped tail.

We had objective evidence for the colour and colour patterns of a dinosaur. The bounds of knowledge had expanded into an area that a week before had been speculation.

Science beats speculation

This is the theme of the story that follows: how science has pushed back speculation in dinosaur science. Not so long ago, the only answers to questions about dinosaurian palaeobiology, such as 'How fast did this dinosaur run? Could this dinosaur crack bones in half? What colour was it?' were little more than guesses, even if informed ones. Now these are questions that can be tested with evidence. That's science, and the switch from speculation to science is a massive advance.

I have had the good fortune to live through this astonishing revolution, starting in about 1970, when the transformation of dinosaurian palaeobiology began. One by one the speculations about evolution, locomotion, feeding, growth, reproduction, physiology, and, finally, colour have fallen to the drive of transformation. A new breed of dinosaur palaeobiologist replaced the older ones, and they have applied a hard eye to the old speculations. Smart lateral thinking, new fossils, and new methods of computation have stormed the field.

Beginnings

Like so many, I became fascinated with dinosaurs when I was young. When I was seven, I was given a classic little book, *Fossils, a Guide to*

Prehistoric Life, by Frank Rhodes, Herbert Zim, and Paul Shaffer. What excited me was that the illustrations were all in colour – unusual still in the 1960s – and that there were not only pictures of fossils, but reconstructions too. The text reflected the knowledge of the time – this is what *Tyrannosaurus* looked like, based on the classic studies by Professor Henry Osborn of the American Museum of Natural History, and this is how the dinosaurs died out, rather slowly, and perhaps as a result of long-term cooling climates (or maybe simply because they were too stupid to adapt to a changing world), according to the ideas of Professor Leigh Van Valen of the University of Chicago.

The assertions were clear, although the only reasons given for why we might wish to accept or reject the explanations was that they were the views of distinguished professors at distinguished addresses (and sometimes with distinguished beards).

Nonetheless, as a seven-year-old, that was all I needed. It never entered my head to question the authority of something written in a book, especially since most of the key information in Rhodes, Zim, and Shaffer was widely repeated. In any case, what could Professors Osborn and Van Valen actually have done in order to test what *Tyrannosaurus* looked like or how the dinosaurs died out? Dinosaurs are long-dead animals, represented now by skeletons and isolated bones. The extinction of the dinosaurs happened 66 million years ago, so how on Earth could a scientist hope to investigate it scientifically?

What is science?

This was the point being made by Sir Ernest Rutherford – the New Zealand-born physicist who made his name at the University of Cambridge with the discovery of the half-life of radioactive elements – when he stated, around 1920, that 'all science is either physics or stamp collecting'. Many hard-nosed physicists might agree with him even today. Nonetheless, he was ruling that much of chemistry, biology, geology, and the applied sciences in medicine and agriculture was not scientific.

I'm sure Rutherford viewed the sciences in a series, reading from left to right from 'strong' to 'weak'. At the strong end were mathematics and physics – his sciences, where experiments are designed and can be repeated with the same outcomes endlessly. These are the sciences where theory consists of equations that can be proved as universal laws, such as gravity or the electromagnetic theory of light. At the other end

Sir Ernest Rutherford, Nobel-prize-winning physicist, and a man
with strong views about what is (and is not) real science.

of the spectrum would be the so-called 'soft sciences' such as sociology,
economics, and psychology.

I expect Rutherford was also thinking about the popularity of
nature among the Victorians, and how the amateur botanists, sea-pool
scourers, and fossil-hunters went out at weekends to collect stuff. Indeed,
collecting specimens for their beauty or for the satisfaction of completing
a list ('I've seen all the birds listed in the handbook') is not science. What
if they were writing down new information, say a new record of a rare
butterfly; that was hardly pushing back the boundaries of science, was it?

What about the historical sciences, such as geology and palaeontology?
They focus on long-past events, such as the origin of the Earth, the
'Cambrian explosion', when so many organisms suddenly appear in the
fossil record, the origin of the dinosaurs, or the origin of humans. These
are singular events that cannot be repeated. Nor can we go back in a time
machine to see what was really going on.

Other historical sciences include archaeology, of course, and physical
geography (the history of climates and landscapes), but also the parts of
astronomy and cosmology that deal with the origin and function of the
universe, and much of biology, which explores the evolution and function

of groups of plants and animals, their ecology and behaviour, as well as unique adaptations and their genetics.

The great philosopher of science Karl Popper gave the answer in 1934, in one of his most important books, *The Logic of Scientific Discovery*. In this, he argued that hypotheses are unlimited, but they must be open to refutation, through his so-called 'hypothetico-deductive method'. Hypotheses can only ever be falsified; they can never be proved. So, if Professor Smith declares, 'My hypothesis is that *Tyrannosaurus* was purple with yellow spots', that is not really a hypothesis because he provided no evidence, and so it can neither be proved nor disproved; it's a belief. (Note, however, we would argue that when we said *Sinosauropteryx* was ginger and had a ginger and white stripy tail, we were doing so scientifically and in a way that could be disproved by another scientist who might fail to find the melanosomes we claimed as evidence.)

In time, Popper explained, the accumulation of evidence corroborates a hypothesis. However, that well-supported hypothesis can then be disproved by a single fact. He gave the example of the swan, once thought – or hypothesized – to be white as a fundamental biological adaptation so they can be camouflaged against the winter snows. But the discovery of a species of black swan – such as the Australian black swan, first encountered by European naturalists in the seventeenth century – disproves the hypothesis, or at least adds a qualification: 'Not all swans are white, and so the camouflage model does not apply to the Australian Black swans.' Popper's key point was that anything that can be set up as a series of testable hypotheses (his hypothetico-deductive method) qualifies as science, and so sociology, economics, psychology, and indeed palaeontology are science if framed correctly.

I have been a little unfair on Rutherford here, as he would have accepted much of what Popper said. He was making a more restrictive claim about general laws. Geologists and biologists have struggled to formulate any universal laws of their subject.

For example: evolution is a universal principle, or set of processes, underlying the entire history of life, as well as modern phenomena such as the evolution of resistance to drugs and pesticides by disease vectors and crop pests. So, evolution is universal and it works, and it provides a vast overarching framework within which thousands of scientists operate throughout their professional lives. But it is not a universal law like gravity or the electromagnetic theory of light; exact predictions cannot be made. Gravity and light are predictable whatever the circumstances, but evolution depends on all sorts of unpredictable factors of organism and environment.

What methods and evidence do palaeontologists use?

When I was a student of biology at the University of Aberdeen in 1976, these concerns were far from my mind. I merely wanted to be a palaeontologist, to be paid (eventually) for doing what I loved – collecting fossils, drawing ancient creatures, and reading about dinosaurs endlessly. We were taught all the subjects in biology, how plants and animals worked, their evolution, ecology, and behaviour.

Then, we had an unusual series of lectures from a professor of the old school – indeed, he probably was not a professor – a wonderfully wrinkled and ancient-seeming man called Phil Orkin. (Checking the University records, I find Phil was born in 1908, and so was sixty-eight when he taught us; he died, aged ninety-six, in 2004, having been at the forefront of leading the small Jewish community in Aberdeen for years.) We were shocked by some of what he told us – that the facts we were learning were probably wrong and would be improved, corrected, and rejected in future.

As students, we struggled with his lectures because they were delivered without notes, and he did not give handouts. Still, Orkin did make us think about what we were being taught – all knowledge is provisional, he told us, and we must strive to make accurate observations. If we eventually made an observation that could overturn an accepted hypothesis, we had better be sure our observation was accurate.

What do palaeontologists have at their disposal? They have fossils, and the rocks from which the fossils come, as well as microscopes to look for fine-scale structures in those fossils – like our melanosomes. They also have methods from engineering, physics, biology, and chemistry to apply to their fossils. Fieldwork supplies great data.

For example, in the 1990s, I was working with Russian colleagues in the red beds of the Permian and Triassic around Orenburg, on the boundary of Europe and Asia. These 'red beds' (so-called because they are beds of rocks such as mudstone and sandstone that are red in colour) extended over hundreds of kilometres, documenting long spans of time through the Permian–Triassic mass extinction, which happened 252 million years ago. As we collected fossils, we also recorded the successions of sediments in detail, and collected samples every metre (3 feet) or so for laboratory analysis. We wanted to find out the geochemistry of the samples, to record the levels of oxygen and carbon throughout the succession in order to give information about the climate and atmosphere, and especially to focus on the time of the great mass extinction, when some 95 per cent of species on Earth died out. We also recorded the orientation of the north magnetic pole through the rock succession, using methods of magnetostratigraphy – from time to time, Earth's magnetism has reversed polarity, so that the north and south magnetic poles have flipped. These crises mark time lines, and so can be used for dating the rocks against a world standard.

The data we were collecting in Russia allow geologists and palaeontologists to test how long the extinction event took, and whether it was one event or many – in fact, there were two bursts of extinction then, separated by 60,000 years. These observations require great care and sophistication in analysis and, together, they provide the essential framework for scientists to explore what caused that catastrophic loss of life and how life recovered (I wrote about this in *When Life Nearly Died*, 2015).

We collected various kinds of fossils in Russia, along the banks of the mighty Ural and Sakmara Rivers, which drain water from the Ural Mountains to the north, and erode down through the Permian and Triassic red beds. The ancient sediments included *body* fossils, skeletons and shells, and *trace* fossils such as tracks and coprolites (fossil faeces). Tracks can show details of soft tissues, such as the pattern of skin on the sole of the foot, and they record the *behaviour* of one or more animals on a particular day 250 million years ago; we can even estimate the speed of the beast from the spacing of its footprints. In Russia, we did not find any exceptionally preserved fossils, showing skin or feathers, for example, but such fossils, as in the case of our Chinese dinosaurs with feathers, can be crucial for palaeobiological interpretations.

Testable methods: bracketing

In making his comments about the feeding behaviour of *Tyrannosaurus*, Professor Osborn referred to modern predators such as lions and hunting dogs. Their behaviour may give us clues about how extinct animals behaved. For example, many hunting dogs today are too small to bring their prey down with a decisive bite to the neck, as a lion or tiger might. So, a small pack of wolves in Canada may follow a moose and snap at its leg tendons trying to break them and so cripple the animal. The wolf could be killed any moment by a kick from the moose, so it has to circle and rush in fast to deliver a bite. After many miles of chase, the exhausted moose may collapse, and the wolves can at last kill it and feast on its flesh. These observations provided ideas for how smaller predatory dinosaurs might have harried their larger prey.

In these cases, the palaeobiologist is using *modern analogues* as a way of making believable, and vivid, his or her assumptions about the fossils. In some cases, study of the modern analogues points the palaeobiologist at things to seek in the fossils. Maybe pack hunting cannot be determined in a dinosaur skeleton, but hunting modes may be deduced from the frequency of broken and damaged bones found in the hunters – were they risk-takers, like modern wild cats and dogs, leaping at larger prey and risking injury?

There is a key question, though: how do you choose your modern analogue? If you are trying to understand the hunting tactics of *Tyrannosaurus rex*, is the wolf, as a mammal, a good analogue? Would an example of hunting behaviour from a lion or eagle, or even from a shark, be equally useful? This question was left unanswered until 1995.

In that year, Larry Witmer argued that an insight he had developed would allow us to say a great deal about every unpreserved detail of, say, *T. rex*. We could describe its eyeball, its tongue, its leg muscles, even its behaviour around egg-laying and hunting. Witmer's insight is called the *extant phylogenetic bracket*. (Phylogeny is the evolutionary history of an organism or organisms.) He reasoned that, if the analogues were well chosen, they could tell us a great deal. For example, in the evolutionary tree, birds and crocodiles are close relatives – they are all archosaurs ('ruling reptiles'), together with the dinosaurs. If crocodiles *and* birds share some detail of the eyeball or the leg muscles, then dinosaurs had it too. We can't say dinosaurs had feathers simply because birds have feathers – crocodiles do not have feathers, so dinosaurs are not bracketed as far as that character is concerned. That's why we can confidently describe the form and function of the eyeball of *T. rex* – not because

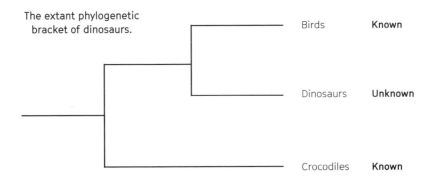

The extant phylogenetic bracket of dinosaurs.

Birds   **Known**

Dinosaurs   **Unknown**

Crocodiles   **Known**

of random comparisons with lions or sharks, but because crocodiles and birds, which bracket the dinosaurs in the evolutionary tree, share most features of eye structure and function. Likewise, we can say that *T. rex* probably showed some minimal parental care after its babies hatched – because crocodiles and birds both share this behaviour.

I can give a concrete example that was crucial in our work on the colour of dinosaur feathers. We studied how colour is expressed in modern bird feathers, and how the different kinds of melanin, the black-brown kind and the ginger kind, are associated with different microscopic organelles. Black-brown melanin is packed into sausage melanosomes, ginger melanin into ball melanosomes. We saw this in bird feathers, and it was always the same. It's also true of all mammals, including humans. In the evolutionary tree, dinosaurs, and most other extinct reptiles, are bracketed by birds and mammals, so this is a universal relationship. Therefore, when we saw the ball melanosomes in *Sinosauropteryx*, we thought, birds have these, mammals have these, and so this works for the bracketed dinosaurs too. *Sinosauropteryx* had ginger feathers.

Testable methods: engineering models

Another testable method in palaeobiology is the engineering analysis of digital models. A digital model is a perfect 3D rendering of an object inside a computer. The model can be rotated and magnified, and the analyst can fly in through the left eye socket of the digital *Tyrannosaurus* skull, and out through its mouth, before returning through the right nostril to explore inside the nasal cavity. The secret to testability is to map the correct material properties onto the bone – in other words, the material properties of bone, as calculated from modern bone: what force it takes to smash a 1-centimetre (⅜-inch) cube of bone, and how far a bone of a certain diameter can be bent before it snaps. Then the engineering analysis can begin.

For her doctoral studies in Cambridge, UK, Emily Rayfield had to work out the form and function of the skull of *Allosaurus*, a large predatory dinosaur of the Late Jurassic. She scanned the skull, and repaired it by replacing missing and damaged bones and removing distortion to create the perfect 3D digital model of the skull. She then assigned material properties to different parts of the skull – hard and brittle for the enamel of the teeth, softer and more pliable for parts of bone around the sides of the skull.

To assign material properties the skull is divided into pyramidal 'cells' or elements, and then a classic engineering method can be applied, *finite element analysis* (FEA). This is the method used by architects and civil engineers to stress-test their designs before beginning construction. Every skyscraper, bridge, or aircraft to which you entrust your life has been pre-tested using FEA.

The argument is that we know the method works. The digital model of a future skyscraper, bridge, or aircraft is stress-tested to see at what point of applied pressure it breaks. This is the basis of the engineering design of the structure before it is completed, and we live in skyscrapers and fly in aircraft designed this way, trusting that the calculations were correct. Therefore, if we use the same approach to study a dinosaur skull or leg, we should accept the results as true. Inside the computer is a perfect functioning model of the extinct animal. This is a pretty amazing claim – that palaeobiology is testable science. Even Ernest Rutherford might have accepted that we can now turn some parts of palaeobiology into rigorous, hard science.

The revolution

I have lived through a revolution. When I started as a student some forty years ago, palaeobiology was a practical subject aimed at solving problems for the oil industry – especially relevant in the town where I grew up, Aberdeen. The granite city was experiencing massive economic growth as a result of the North Sea oil boom. If my professors talked about form and function or evolution, they did so a little apologetically, because they were straying from hard facts.

Through my scientific career, I have seen dinosaur science (and palaeobiology in general) change from natural history to testable science. New technologies have revealed secrets locked in the bones – we can now work out the colour of dinosaurs, their bite forces, speeds, and levels of parental care. I have taken an active part in the debates about reconstructing the tree of life, the Jurassic Park phenomenon and the viability (or not) of dinosaur DNA, the CT scanning and digital imaging revolution, and new engineering models to test the bite force and running speed of *T. rex*, as well as the colour of dinosaurs.

Much of the press coverage of modern palaeobiology focuses on remarkable new fossil finds, such as giant sauropod dinosaur skeletons from Patagonia, dinosaurs with feathers from China, and even a tiny

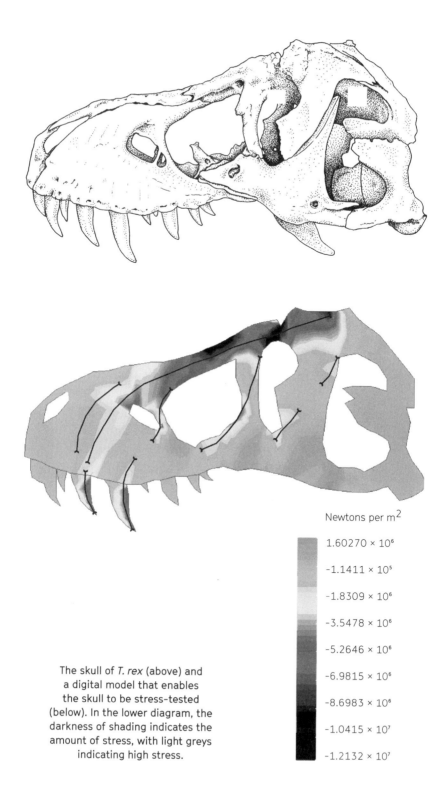

Newtons per m^2

1.60270 × 10^6

-1.1411 × 10^5

-1.8309 × 10^6

-3.5478 × 10^6

-5.2646 × 10^6

-6.9815 × 10^6

-8.6983 × 10^6

-1.0415 × 10^7

-1.2132 × 10^7

The skull of *T. rex* (above) and a digital model that enables the skull to be stress-tested (below). In the lower diagram, the darkness of shading indicates the amount of stress, with light greys indicating high stress.

dinosaur tail in Burmese amber. New fossils are the lifeblood of modern palaeobiology, of course, but it is the advances in technologies and methods that have driven the revolution in scope and confidence.

The aim of this book is to show all the latest amazing fossils, and to take the reader behind the scenes on the expeditions and in the museum laboratories. The key theme throughout is the transformation of a historical science from its roots in Victorian natural history to a highly technical, computational, and thoroughly scientific field today. These have been exciting times of rapid change and astonishing new discoveries, happening at a rate never seen before.

Chapter 1

Origin of the Dinosaurs

One thing is known for sure: the dinosaurs originated during the Triassic period, between 252 and 201 million years ago. Nearly everything else is uncertain. For example, just when did they originate, in the Early or Late Triassic? What was the world like as they emerged on the scene? Did they force their way to dominance of global ecosystems by fighting hard for their place against other beasts, or did they achieve their position by good luck? When I began my career as a palaeontologist, back in the 1980s, these were all hot topics of discussion. Solving the questions has been my life's work, but I can't say everything is sorted out: whenever one problem is resolved, further questions are raised. It's a story of changing ideas about evolution, new fossils, and new analyses.

As part of my doctoral studies I tried to work out an ecological model for the origin of the dinosaurs. The 'standard' model then was a three-step process. First, the synapsids, ancestors of mammals, were the key herbivores and carnivores. Then, the synapsids were replaced

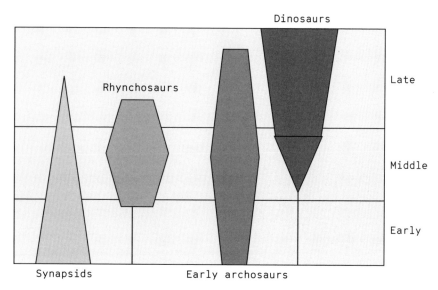

The classic model for dinosaur origins by progressive competitive replacements in the Triassic.

by rhynchosaurs as herbivores, and early archosaurs as carnivores. Archosaurs include birds and crocodiles today, and dinosaurs and their ancestors. Finally, the rhynchosaurs and early archosaurs gave way to dinosaurs. We will encounter all these animals shortly, especially the rhynchosaurs and the first dinosaurs.

These three steps were said to form an ecological relay, in which one group gives way to another, which in turn gives way to another. This ecological-relay model for dinosaur origins had been presented by the two great American palaeontologists at the time, Al Romer and Ned Colbert, who were the authors of all the standard textbooks, so their ideas were widely distributed and widely read. Importantly, the Romer–Colbert relay model assumes competition between all these animals, and that the dinosaurs in some way fought their way through to dominance. How did they do it? Possibly because they had erect or upright posture, and so could run faster than their unsuccessful neighbours. In broader evolutionary terms, the Romer–Colbert ecological-relay model was firmly framed within an assumption that large-scale evolution was progressive.

I presented an entirely opposing view in a 1983 publication, as a cheeky young research fellow. I argued that the dinosaurs had exploded onto the scene about 230 million years ago, not after a long competitive struggle, but following an extinction event. The rhynchosaurs and early archosaurs had been killed off by climate change, which had led to drying conditions and the prevalence of different kinds of plants, notably conifers. The rhynchosaurs chomped unhappily at the unforgiving needles and cones of the new arid-land conifers; they were, in fact, adapted to feeding on equally tough, but more nutritious, vegetation such as seed ferns, but these plants required damper climates. Perhaps the drying climate and spread of conifers led to the rapid demise of the seed ferns, and then of the rhynchosaurs. In their heyday, the rhynchosaurs had been hugely abundant, making up as much as 80 per cent of the entire fauna. After their extinction, the dinosaurs took their chance and expanded into empty ecospace – this, I argued back in 1983, is opportunism, rather than progress.

This new idea of mine was probably quite annoying for the established palaeontologists. Indeed, I had a somewhat heated, and unexpected, discussion with the doyen of Triassic dinosaur studies in Britain, Dr Alan Charig, head of the dinosaur section at the Natural History Museum in London. He buttonholed me at a conference in Manchester in 1985 and we had a serious discussion – in the showers. (In those days conferences were typically housed in university halls of residence with communal shower facilities.) I was trying to convince Charig that we should use numerical, phylogenetic methods to resolve big questions in macroevolution, but he could not agree; we agreed to differ, and parted on good, if slightly damp, terms.

This is a story, then, of evolution on the large scale, but it depends also on a good knowledge of the fossils, the rocks, and the models of large-scale evolution. We shall look at the ecology of Triassic beasts, then the rhynchosaurs (an odd, but endearing, group of Triassic animals that are key in many ways), then the question of the very first dinosaurs, and how we can put together the story of fossils, changing climates, and mass extinctions to see how dinosaurs rose to dominate the Earth.

Ecology and the origin of dinosaurs

So why did Romer, Colbert, and Charig argue that dinosaurs outcompeted their rivals? It was partly that progress was assumed in evolution – dinosaurs replaced their inferior competitors (the synapsids, rhynchosaurs, and early archosaurs), and were in turn replaced by mammals, 180 million years later. Each step along the way marked an improvement of some sort, by which the animals became faster, smarter, or at least better competitors.

This is in some ways pure Darwin – survival of the fittest, constant improvement. We have learned since 1980, however, that evolution is not unidirectional or relentless. In fact, the physical environment keeps changing, as, for example, climates become warmer or cooler, continents move positions, mountain ranges emerge, and sea levels rise and fall. As conditions change, the plants and animals embedded in them keep adapting in a purely Darwinian evolutionary way, but they never quite attain perfection. Environmental changes are unpredictable, and somewhat random, so species are on the whole good at what they do, but probably never perfect.

The focus in the 1980s was on posture. Today, reptiles such as turtles, lizards, and crocodiles are sprawlers: that is, they hold their arms and legs quite a bit out to the side. When they walk, if you view them from above, each arm and leg describes a wide arc, and the backbone bends from side to side. Sprawling reptiles keep their belly close to the ground, and can generally only scuttle fast for short distances. Mammals, on the other hand, have erect gait, meaning their arms and legs are tucked right under their body. When they walk they use the whole length of the arm and leg

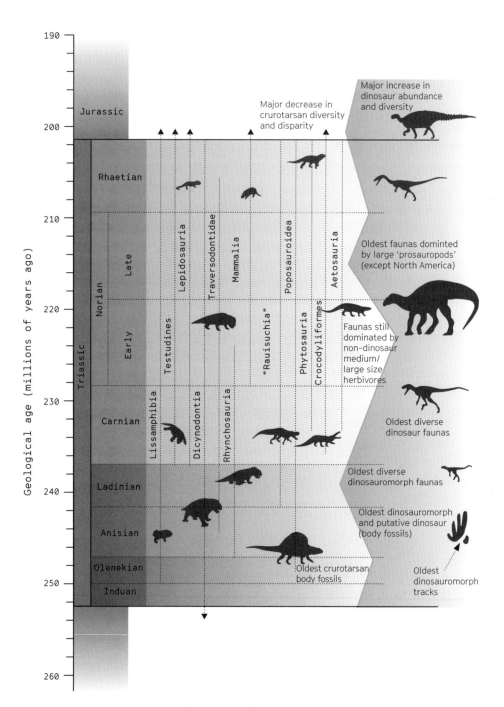

Key stages in the origin of dinosaurs
through the Triassic.

in making a stride, and there isn't much lateral movement of the limbs or the body. Famously, many mammals, such as horses or wolves, can run fast for very long distances, which generally sprawlers cannot do.

The argument, then, was that there had been a major transition in the posture of reptiles during the Triassic. The synapsids and rhynchosaurs were mainly sprawlers, so it was argued, and the dinosaurs were erect, and this gave them the competitive edge. The dinosaurs lived life at a faster pace than their synapsid and rhynchosaur precursors, and they won out in a kind of biological arms race that lasted through the entire 50 million years of the Triassic.

This theory seemed clear and it explained the data. However, I found it unsatisfactory, and this came from my realization that the fossils and rocks told a different story. The dinosaur takeover was rapid, not gradual, and there was no evidence for direct competition. This grew out of my doctoral studies on rhynchosaurs, a group of reptiles that were ecologically dominant worldwide just before the dinosaur explosion.

Rhynchosaurs

Starting my doctoral studies in 1978, I was assigned the topic of working on *Hyperodapedon* (see overleaf), a rhynchosaur from the Late Triassic of Elgin in Scotland, by my supervisor, Alick D. Walker, at the University of Newcastle-upon-Tyne. My job was to examine the twenty or so specimens of this rather odd, four-legged, bulky herbivorous reptile. The specimens had been collected since the 1850s in yellow sandstones around Elgin, an attractive market town in northeast Scotland.

The fossils were annoying because they were holes in the rock. At some point in the 230-million-year history of that corner of Scotland, the rocks had been buried, squeezed, baked, and then uplifted. The bone material was still there, but awkwardly putty-like. In Victorian times, the museum preparators had laboured hard with hammer and chisel to remove the fine-grained sandstone from these squishy bones, but the results were generally disappointing.

Alick Walker had had the insight in the 1950s, when he began his life's work on the Triassic fauna near Elgin, to remove any remaining bone scrap and then make high-fidelity casts from the natural rock moulds. By some means that I never discovered, he selected PVC as his moulding material of choice. This is the stuff of rubber gloves, starting out as a thick liquid that can carry colour, which is poured into a mould, baked to cure, and then pulled out again. The extreme flexibility and strength of a PVC rubber glove was what we wanted – after pouring and baking, the PVC had infiltrated deep into every cavity and crack within the rock.

Sometimes, I had to round up three or four fellow students to help me haul a PVC cast of a leg bone or a skull out of the rock. It was worth it, though, because the sandstone retained very fine details, revealing, for example, the tear duct of the eye, major blood vessels, and bone sutures in the skull of *Hyperodapedon*.

Now, rhynchosaurs could be up to 1.5 metres (5 feet) long, and they have a very recognizable skull, with a hooked snout, a sort of grin when viewed in side view, and a very broad skull at the back. The breadth of the skull created a huge space between the (small) braincase in the middle of the skull and the jaws, which in life was filled with several powerful jaw muscles. The diameter of a muscle gives a measure of its power, and there was no doubt that rhynchosaurs had amazingly powerful jaws. Their dentition bore this out, consisting of several rows of teeth, being

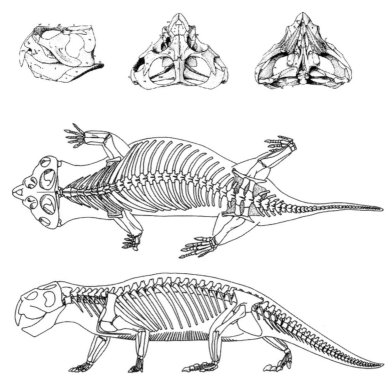

The rhynchosaur *Hyperodapedon* from Elgin
in Scotland – a page from my doctoral thesis.

Genus:	*Hyperodapedon*
Species:	*gordoni*

Named by:	Thomas Huxley, 1859
Age:	Late Triassic, 237–227 million years ago
Fossil location:	Scotland
Classification:	Archosauromorpha: Rhynchosauria
Length:	1.3 m (4¼ ft)
Weight:	50 kg (110 lbs)
Little-known fact:	*Hyperodapedon* lived worldwide and is known from Argentina, Brazil, India, and Tanzania.

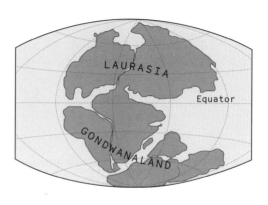

emplaced at the back of each jawbone, and expanding the tooth row as the animal grew larger. Teeth near the front were worn flat by close wear against those of the opposite jaw. Indeed, one of the first palaeontologists to describe rhynchosaurs, the famous Victorian supporter of Darwin, Thomas Henry Huxley, compared their jaw action to the closing of a penknife – the lower jaw is the blade, and it fits snugly into the deeply grooved upper jaw. This shows that the only jaw action they were capable of was to cut the food precisely, as if with a strong pair of fabric scissors, an action technically termed 'shearing'. The jaws could not move sideways, so they could not chew their food.

Understanding the adaptations and the world of the rhynchosaurs was important, as they had been the dominant plant-eaters before the dinosaurs exploded onto the scene. How quickly were they replaced, and were they pushed out by the dinosaurs, or did they simply die out for other reasons?

As I came to finish my doctorate, I faced a dilemma. Rhynchosaurs were lovely, or so I thought, with their happy smiles and precise shearing jaw action, but they were all the same. Hundreds of skeletons of rhynchosaurs had been found, not only in Scotland, but also in Triassic rocks of similar age in Brazil, Argentina, India, Tanzania, Zimbabwe, Canada, and the United States. At first, these different finds accumulated several different names, but they have all been re-studied, by me and others, and we struggled to highlight any differences. The rhynchosaur *Hyperodapedon* lived worldwide in the Late Triassic at the same time as what were then the world's oldest dinosaurs.

What was the first dinosaur?

Up to the year 2000, the oldest known dinosaurs were all from the Late Triassic, and dated at about 230 million years ago. The oldest decent dinosaur specimens were found in the Ischigualasto Formation of Argentina in the 1950s and 1960s, when Al Romer from Harvard and local Argentinian geologists began excavating. The Ischigualasto country lies in sight of the Andes Mountains, and has been uplifted on the flanks of that great mountain range. Geologists travelled 200 kilometres (124 miles) north of the regional city of Mendoza in San Juan Province, first on passable roads, and then on dirt tracks as they got close to the dinosaur sites. The Ischigualasto landscape consists of wide, bare valleys, eroded by seasonal floods tearing down the eastern flank of the Andes, and exposing broad strips of badland scenery, with sharp ravines cut in the rock, revealing the mix of red- and grey-coloured sandstones. The fossil beds are in the Ischigualasto Provincial Park, located in the romantically named Valle de la Luna, the Valley of the Moon. Picking over these barren landscapes is hard work, but it's ideal fossil-hunting territory as there is no soil or vegetation: the fossils stand out as white and purplish bones in the rock.

The fossil collections made by Romer went back to Harvard and he and his students published a series of papers describing the fossils, including the dinosaur **Herrerasaurus** (see overleaf). This was named by Osvaldo Reig, a famed Argentinian palaeontologist, in 1963. *Herrerasaurus* was a large animal, some 6 metres (20 feet) long, with great meat-cutting jaws. It was a biped, clearly capable of fast movement on its powerful, upright hind legs, each equipped with broadly spreading toes. It had long, powerful arms, and could have used these to grab prey. Its jaws were lined with twenty-five scimitar-like teeth, each with a serrated edge, like a steak knife. It pains me to report that *Herrerasaurus* was probably large enough to feed on the most abundant animals of its day, the rhynchosaurs. Other animals in the Ischigualasto rocks include smaller animals such as the dinosaurs *Eoraptor* and *Panphagia*, each about 1 metre (3 feet) long, as well as armoured plant-eating early archosaurs such as the aetosaurs, and some smaller carnivorous synapsids that probably looked like partly hairy, partly bald rats.

Scientific expeditions to Ischigualasto Provincial Park in the 1990s revealed dozens more dinosaur skeletons, including fairly complete skeletons of *Herrerasaurus* and *Eoraptor*. The Ischigualasto dinosaurs, dated as about 230 million years old, are matched by smaller faunas of similar dinosaurs from rock formations of the same age in Brazil, India, and North America, which is why I took them as the marker of an explosive worldwide diversification of dinosaurs following a major environmental crisis.

Then, after the year 2000, a series of new discoveries rather suddenly pushed the date of the origin of dinosaurs back by 15 million years, and placed it in an entirely new and unexpected context.

The first hint of the revolution in our understanding came from Poland. In 2003, Jerzy Dzik, Director of the Palaeontological Institute in Warsaw, reported a skinny reptile from southern Poland, called **Silesaurus** (see overleaf). The fossil was remarkably complete, some 2 metres (6½ feet) in length, with a long, slender body, long, thin arms and legs definitely held in the erect posture, and a long neck and sleek head.

Genus:	**Herrerasaurus**
Species:	*ischigualastensis*

Genus:	**Silesaurus**
Species:	*opolensis*

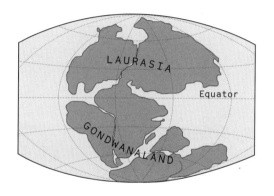

Named by:	**Osvaldo Reig, 1963**
Age:	**Late Triassic, 237–227 million years ago**
Fossil location:	**Argentina**
Classification:	**Dinosauria: Saurischia: Herrerasauridae**
Length:	**6 m (20 ft)**
Weight:	**270 kg (595 lbs)**
Little-known fact:	***Herrerasaurus* looks like a theropod, but appears to be an early saurischian, neither theropod nor sauropodomorph.**

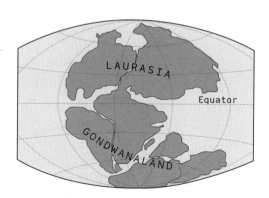

Named by:	**Jerzy Dzik, 2003**
Age:	**Late Triassic, 227–201 million years ago**
Fossil location:	**Poland**
Classification:	**Dinosauromorpha: Silesauridae**
Length:	**2.3 m (7½ ft)**
Weight:	**40 kg (88 lbs)**
Little-known fact:	**The fossils come from a clay pit used by a cement company.**

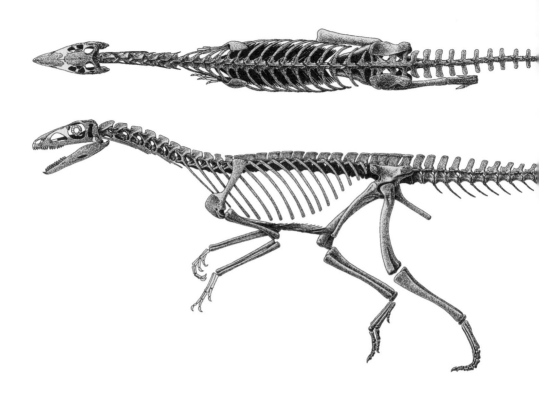

It looked as if it mainly ran as a biped, and could then use its long, slender arms to walk slowly on all fours. The jaws are lined with peg-like teeth, and there is a bony lip at the front of the jaws, so it seems that *Silesaurus* was a plant-eater that nipped at leaves with the bony tips of its jaws, and chopped the food further back in the jaws. *Silesaurus* sort of looks like a dinosaur, but not quite. Could it point back to the ancestry of dinosaurs?

The second Polish surprise came in 2011, when fossilized slender, three-toed footprints from several localities were reported as definitely dinosaurian by Steve Brusatte, Grzegorz Niedźwiedzki, and Richard Butler. Their discovery was disputed – can we be sure these little footprints were really made by a dinosaur, or could they perhaps have been made by something like a dinosaur, maybe even a silesaurid? Well, yes, possibly, but in a way it doesn't matter.

The clincher came in 2010, when Sterling Nesbitt reported a Middle Triassic silesaurid from the Manda Formation of Tanzania, called *Asilisaurus*. The Manda Formation consists of red-coloured sandstones laid down in ancient rivers, and the rocks are now exposed under the thin soil and the burning hot sun of southwestern Tanzania, near the shores of Lake Malawi. The first fossils were found there 100 years ago,

Silesaurus, a member of the group that is closest to the dinosaurs in evolutionary terms.

but renewed explorations by Sterling Nesbitt and his team have revealed many remarkable new specimens.

The discovery of *Asilisaurus* unequivocally re-dated the origin of dinosaurs back from 230 to 245 million years ago, or older. The point was that the Polish slender dinosaur-like animal *Silesaurus* was not alone. In fact, it turns out that *Silesaurus* is the exemplar of a whole new group, named the Silesauridae in 2010. Half a dozen little animals from the Middle and Late Triassic of South and North America were assigned to this group...and then along came the oldest silesaurid, *Asilisaurus*. All these little animals looked somewhat like dinosaurs because it turned out that the Silesauridae were the nearest relatives of the Dinosauria (the formal name for dinosaurs, invented in 1842, as we shall see in Chapter 2), meaning they shared an immediate common ancestor. If the Silesauridae had originated by 245 million years ago, then their immediate relatives, the Dinosauria, must have too. There is even a possible dinosaur, *Nyasasaurus*, from the Manda Formation, but it is known only from isolated bones.

The macroecology of dinosaur origins

If the dinosaurs originated in the Early Triassic rather than the Late Triassic, then this shifts their time of origin back into one of the most turbulent periods in the history of life. This was when life was recovering from near-complete annihilation, and Earth environments were convulsed repeatedly by terrible episodes of acid rain, global warming, and loss of oxygen from the ocean floors. It all started 252 million years ago during the largest mass extinction of all time: the Permian–Triassic mass extinction.

The extinction had been triggered by huge volcanic eruptions in what is now Siberia, which drove an episode of profound environmental destruction – acid rain and extreme warming on land destroyed the forests, and plants and soil were washed into the sea, leaving gaunt, rocky, and baked landscapes. Shallow seas were swamped with the debris and acid and warming, and this perturbed normal ocean cycles. Life on land and in the sea was destroyed, and only 5 per cent of species survived.

Normally, after a mass extinction event, life can recover in a reasonably benign world. The Early Triassic world was far from benign, however. For 6 million years after the crisis, repeated paroxysms of eruptions and environmental destruction occurred. Life would recover for half a million years, and then it would be set back again. Into this perturbed world came the first dinosaurs, taking their chance against other groups in the bleak post-extinction environment.

In my 1983 paper, I had challenged the competitive model for dinosaur success and suggested an alternative extinction model. To test this hypothesis, I had documented occurrences of fossils, identified what they were, and matched them to geological time as best I could. This at least allowed me to test the pattern of change. My data showed that there had been a pretty sharp change in the composition of reptile faunas through the Triassic. Romer, Colbert, and Charig had been right that we started the Triassic with faunas of synapsids, we passed through faunas dominated by rhynchosaurs, and ended the Triassic with dinosaurs everywhere. But the change happened quickly, in one event, some time around 230 million years ago.

My model was explicitly ecological. This meant I didn't simply record the presence or absence of different species, but I also wanted to document how ecologically important they were. This required some knowledge of their size and likely diets, but also of how abundant they were. In other words, in any particular location, out of 100 specimens, how many belonged to each group? I found to my surprise that, when there were rhynchosaurs, they typically made up 50 per cent or more of the fauna. Indeed, at Elgin, and in some other locations, they could represent 80 per cent or more of all specimens.

This attempt to represent ecological importance in terms of relative abundances showed that rhynchosaurs were the dominant herbivores all round the world, and then at a particular point, 230 million years ago, they disappeared. This was the point. They didn't dwindle from 80 per cent to 40 per cent to 20 per cent as you passed up through the rock sections. One minute they were there, and then a few metres higher in the rock section they were gone. It might have been recorded simply as the loss of one or two species worldwide, but ecologically those few species of rhynchosaurs had dominated their ecosystems, and their loss must have left a large gap. (We shall return later to a likely reason as to why the rhynchosaurs suddenly vanished.)

That was the crux of my argument in 1983. Sudden death of the dominants, ecologically speaking, followed by the rise of the dinosaurs. The dinosaurs were already there, as we have seen in the Ischigualasto Formation, and they were diverse and important, but they made up only 5–10 per cent of the fauna. After the disappearance of the rhynchosaurs, dinosaurs switched from representing 5–10 per cent of their faunas to more than 50 per cent in many parts of the world.

New methods and new models

Advances in palaeontology don't all come from new fossils. There are also advances in computational methods and, although they might seem less exciting because they do not involve Land Rovers, sweat, and exotic desert locations, they can be crucial in resolving questions.

When I did my early studies on Triassic ecosystems, the numerical tools were quite feeble. I could only use simple statistics, such as counting up proportions of specimens to describe what was going on. Now, we have access to new mathematical methods that were developed to allow biologists to make comparisons between modern species while properly taking account of their evolutionary relationships. They also allow biologists to work out the so-called ancestral state of any character or trait, which can be a physical feature, such as body size or leg length, or a behavioural feature, such as egg clutch size or feeding ritual. By plotting the known data onto an evolutionary tree, they can estimate down the tree what the ancestors would have been like, and then use this knowledge of calculated ancestral states to look at rates and types of change through geological time.

In an example of the application of these new numerical methods, the Colbert–Romer model for Triassic ecological relay between synapsids and archosaurs (including dinosaurs) could be explored. Roland Sookias did this as part of his doctoral work in 2012. He documented the body size of several hundred synapsids and archosauromorphs (archosaurs, rhynchosaurs and relatives), and tracked the changing sizes through time. He found that archosauromorphs became larger through the Triassic,

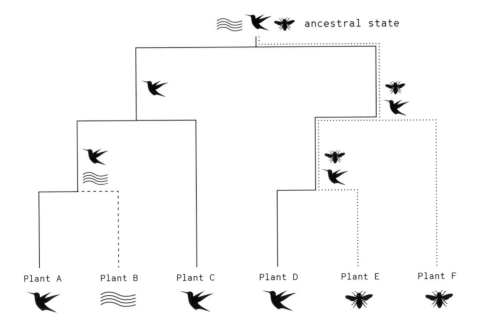

Working out the ancestral states of plants that are
pollinated by insects, birds, or the wind.

mainly represented by the dinosaurs, and synapsids became smaller
as they shrank down to the tiny shrew-like mammals at the end of the
Triassic. But was this a driven trend or just passive change?

Sookias was able to fit different models of evolution to his data,
and mostly it seemed that evolution was simply proceeding in a random
way. That is, the changes in body size were surely happening, but with
sufficient local variations, and sufficiently slowly, that it was not possible
to say that the changes in body size were *driven* by some powerful
evolutionary force. If the size change through the Triassic had been a
driven trend, it would have been possible to make a case that natural
selection was in play, and there was selective pressure for larger size.
All Sookias could say was that the synapsids were getting smaller, on
the whole, and the archosaurs were getting larger, but they could have
been changing size in more or less random ways. Therefore, this did not
disprove the Romer–Colbert model in which dinosaurs outcompeted
their precursors, but it didn't lend the idea any support either.

In an earlier study, Steve Brusatte, then a Masters student in Bristol,
studied the same question, but looking more closely at the first dinosaurs
and the early archosaurs they replaced. He decided to measure rates of
evolution not just from one trait, such as body size, but from all aspects

of their anatomy. He made a huge table of 500 traits for each beast, and used standard statistical methods to reduce this great mass of data into something more manageable.

One way of visualizing such huge and complex data sets is to seek some main directions of variation, and extract these and plot them up as a so-called morphospace, meaning literally 'shape space'. A morphospace is a graphical way to show how the morphology, or external appearance and physical characters, of organisms vary, and it has the great strength of summarizing huge amounts of information into something we can more easily understand. Species that are most similar plot out close together, and those that are most different in life are plotted far apart in the morphospace.

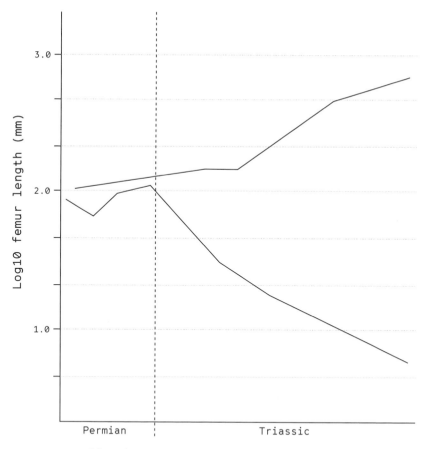

While archosauromorphs (top line) became larger during the Triassic period, synapsids (lower line) became smaller.

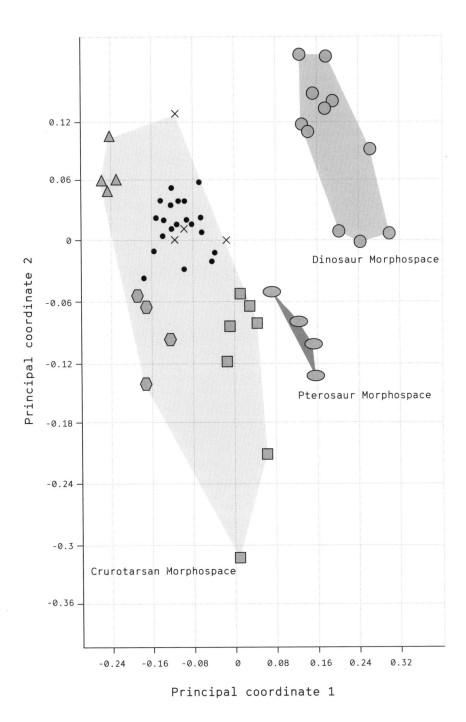

A diagram representing the morphospace of dinosaurs and
other Triassic reptiles, corresponding to their adaptations.

Chapter 1

The morphospace for the dinosaurs and early archosaurs showed areas of morphology occupied by each group. They did not overlap – possibly, but not conclusively, suggesting they were not directly competing. When Brusatte extracted rates of change, he found that dinosaur morphological variation expanded as the group diversified in the Late Triassic, but so too did that of their supposed competitors, the crocodile-like archosaurs, or crurotarsans. There was no sign that the explosion of dinosaurs was hitting the other archosaurs and forcing them into retreat. In fact, they all seemed to be diversifying and occupying new morphospace, in parallel; far from being crushed by the new dinosaurs, the other archosaurs were apparently flourishing. In their ecosystems, the new dinosaurs such as *Herrerasaurus* and *Plateosaurus* were still preyed upon by some of their crurotarsan cousins.

These numerical studies are part of a new wave of computational research in macroevolution. The methods arc tough to learn, but my students absorb them like the air they breathe. These methods have opened the floodgates to enable new explorations of the various phases of the evolution of dinosaurs, and how different groups came and went, and are providing ways to tackle big evolutionary questions that were thought impossible to answer when I began to work on these questions back in the 1980s.

The origin of dinosaurs as a three-step process

So, where do the new discoveries of much older dinosaurs and the new computational studies leave us? Do we have a proper understanding of the dynamics of the origin of the dinosaurs? Who was right –was the Romer–Colbert–Charig ecological-relay model, with a drawn-out, competitive rise of dinosaurs correct? Or was my 1983 mass extinction-opportunism model correct?

In fact, we all got it wrong in different ways. I was wrong to assert that dinosaurs had emerged and exploded in the Late Triassic, as we now know their first steps took place 245 million years ago, in the Early and Middle Triassic. Romer and Colbert were right that the origin and initial rise of the dinosaurs lasted throughout some 40 million years of the Triassic, although they had not based this on any knowledge of more ancient dinosaur specimens.

The erect posture of dinosaurs was clearly key to their success, again as Romer and Colbert had said, but in a somewhat different sense from the one they had argued. In other words, if the first dinosaurs arose much earlier in the Triassic than anyone had known back in the 1970s and 1980s, then they did not out-compete the synapsids, early archosaurs, rhynchosaurs, or any other groups. In evolution, organisms mostly avoid competition by shifting their ecological niches – they choose a different diet or geographic range. Discretion is the better part of valour; successful organisms live to fight another day. Evolution probably isn't exactly 'red in tooth and claw', as Alfred Lord Tennyson suggested; more pinkish in tooth and claw.

What set off the big burst, though, the second stage? If this hadn't happened, dinosaurs might have continued to be rather rare animals, maybe 10 per cent of their faunas. I might have got it right back in 1983. However, new evidence is showing an ever-closer link between the dinosaur explosion and the so-called Carnian Pluvial Episode. This event had been noted and named by Mike Simms and Alistair Ruffell back in 1989. They had observed in Late Triassic rock sections around the UK, and other parts of Europe, that something unusual was happening. The generally dry climates were interrupted by a pluvial phase, when rainfall increased substantially, and then conditions returned to dry. The evidence for the climate change came from the rocks themselves and from the plant remains, which can be classified as wet-loving (mosses, liverworts, horsetails) or dry-loving (conifers).

There things rested until 2012. The dinosaur explosion–Carnian Pluvial Episode link was emphasized from time to time by me, by Simms, and by Ruffell, but nobody else paid much attention. Then, independent work published in 2012 by Italian geologist Jacopo Dal Corso changed everything. Dal Corso found that the rocks documenting the Carnian Pluvial Episode reflected the results of major volcanic eruptions in western North America. There, about 232 million years ago, great eruptions produced huge volumes of volcanic lava, the Wrangellia basalts, seen today around Vancouver and northwards along the coast of British Columbia.

Dal Corso argued that the eruptions had been so huge that they had caused a shock climate change worldwide. Just as at the end of the Permian 252 million years ago, huge volumes of carbon dioxide pumped out of the volcanic vents caused global warming, as well as acid rain. The warming and acid rain killed life on land, and led to ocean acidification and loss of oxygen in bottom waters, and Dal Corso noted extensive evidence of extinctions in the sea in localities across Europe and North America. The warming also led to mega-monsoonal conditions around

Middle Norian-Rhaetian
~225-202 Mya

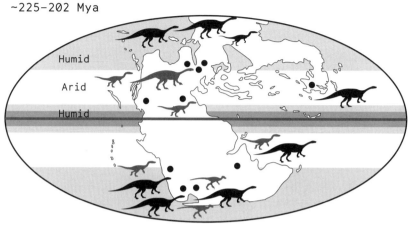

Humid

Arid

Humid

Late Carnian-early Norian
~232-223 Mya

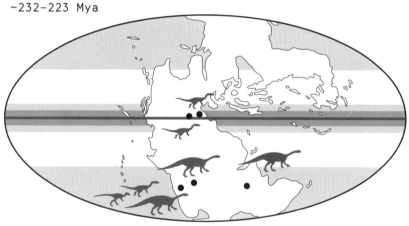

Common		
Rare		
Ornithischia	Sauropodomorpha	Theropoda

Changing climate belts in the Late Triassic, and the
movement of dinosaurs from the southern continents
to the whole world in the later period.

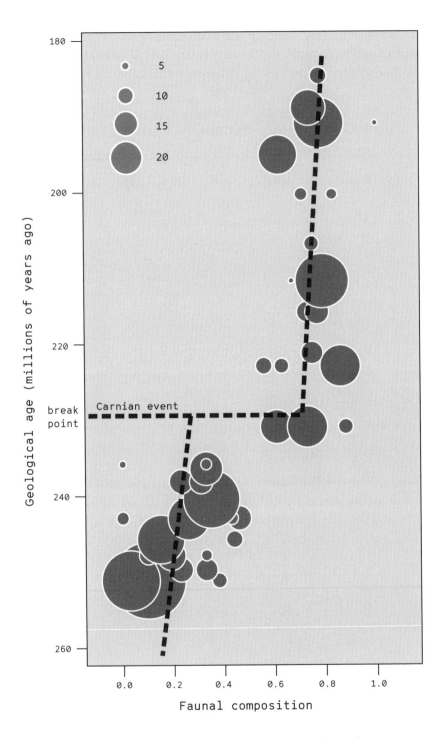

The break point in ecosystem evolution was triggered
by the Carnian Pluvial Episode 232 million years ago.

the broad equatorial belt, which at that time encompassed all the dinosaur localities in North and South America, Europe, and India. After the eruptions ceased, conditions returned to hot and dry, and this was the killer.

These new studies stimulated me to look again at the ecological data I had collected back in 1983. I was aided by students Cormac Kinsella and Massimo Bernardi. We checked all the data, and improved the sample of specimens to over 7,700, and worked out the per centages of all the key groups – rhynchosaurs, dinosaurs, and the others. Then, I plotted the ecological ratios as a so-called bubble plot, where each bubble represents a different fauna, with the centre dated correctly according to the geological time scale. The size of each bubble represents the number of specimens in the sample, and I plotted each one against the value of a ratio that essentially gives us a measure of how much dinosaurs dominated the faunas.

There did seem to be a step up from values around 20 per cent to values around 70 per cent, but such a guestimate wasn't good enough. Critics might say we were just imagining things. So, we applied a numerical method called 'breakpoint analysis' to the data. This method attempts to calculate the line that best explains all the data, but you allow it to make one (or more) breaks. We programmed the model to make one break, and set the calculations running. After a short time, the answer came back – the best-fitting line had a break at exactly 232 million years ago.

This we took as independent evidence that something happened at this time that fundamentally reset the nature of Triassic reptile ecosystems. Dinosaurs had indeed originated long before the Carnian Pluvial Episode, but they had not succeeded in taking over the faunas – in fact they continued to do what they had done before, and would have seemed inconsequential to any observer. The key jump happened 232 million years ago, and the new study provided a marker that linked the ecological revolution when dinosaurs exploded in significance to the environmental upheavals in the middle of the Carnian.

Cementing the link between the Wrangellia eruptions and the Carnian Pluvial Episode was a powerful idea, and it confirmed the environmental shock that perhaps killed off the rhynchosaurs and other dominant animals, so providing the opportunity for dinosaurs to diversify explosively into empty ecospace. Still, the dating needed to be right, and, usefully, there had been step-change improvements in our understanding in recent years.

Dating dinosaurian diversification

So far, I have been quoting geological ages (see Timeline, pp. 6–7) such as 230, 232, and 245 million years ago. Geologists do that. But how do we know? This is crucial for all hypotheses we might have about dinosaurian origins and their eventual disappearance. We have to be able to set the time of ancient events in terms of millions of years, and also match or correlate the rocks from continent to continent to test whether a dramatic event in, say, Argentina is matched by a similar-looking crisis in the Italian rock sections.

Dating the rocks is a core activity by geologists. The roots of the science of dating the rocks, called stratigraphy, were profoundly practical. When the humble English surveyor William Smith began his efforts in the 1790s, he was self-taught. He was one of the first economic geologists and he was paid for results. At that time, there was no evidence to suggest that the rocks under our feet were anything other than a complete jumble. The idea of a geological map that showed the orderly arrangement of different rock formations, and of a time scale that placed them in sequence and could be used across various locations, was unheard of. Smith laboured through his life to establish both principles.

Smith practised in the early years of the Industrial Revolution in Britain, when every land owner was delving for coal – often completely at random. The reasoning was something like this: my neighbour Arkwright finds coal at twenty yards beneath his fields, so I should also find coal at the same depth. Sometimes it worked, and sometimes not. If the two spots were separated by a fault, the rock succession could be entirely different. Smith used his skills in mapping and stratigraphy to work wonders: he could tell people where to dig, and – importantly – where not to dig. If the rocks were dated as Jurassic, there might be coal below because the Jurassic period is younger than the Carboniferous, whose forests and swamps became the coal seams of the future. If your neighbour's rocks were Silurian in age, however, then there would be no coal; the Silurian is older than the Carboniferous. The Jurassic and Silurian rocks might both be rather similar-looking, dark grey limestones, but the fossil content told Smith the age, and he could then turn this knowledge into cash.

Since Smith's day, and thanks to efforts in every land, the geological time scale with its main divisions, the geological eras and periods, were all more or less named by 1840. The geological time scale worked everywhere, and correlations were made regionally by walking across country and mapping, and then internationally by comparing fossils. Smith's Jurassic faunas of ammonites and bivalves could be matched around the world. This marked a line of equivalent age from England to France to Russia to Argentina...Such efforts in the identification of rock ages by fossil assemblages are just as commercially valuable today as in Smith's day, now especially in the oil industry, where companies spend billions of dollars drilling. They want to know ahead of time what they are going to be drilling through, and whether they have to drill 50 metres or 5 kilometres to reach oil.

Stratigraphy using fossils does not give exact ages. These come from radioisotopic dating. Soon after the discovery of radioactivity in the 1890s, the Nobel-prize-winning physicist Ernest Rutherford suggested in 1905 that radioactive decay could provide an exact chronometer for dating rocks, and so for dating the origin of the Earth, which then points to the date of the origin of the universe. The eager young geologist Arthur Holmes seized on the idea, and he had compiled a list of key dates by 1911, at the age of twenty-one. Since 1911, radioisotopic dating has become an important part of laboratory-based geology, with ever more powerful mass spectrometers being deployed. A great strength of the approach is that the same rock can be dated by different means and in different laboratories to cross-check the estimates.

There is an extensive international endeavour to improve precision (tightness of the estimate; size of error bars) and accuracy (is it right?) of exact rock dates, and the standard geological time scale is revised in detail every few months, as dates are tuned to be sharper and sharper, and more comparable. When I started my studies of geology in the 1970s, we were told to assume an error of plus or minus 5 per cent on any radioisotopic date. In some cases now, precision has improved a hundredfold, to an error of plus or minus 0.05 per cent. So, the dating of the Carnian Pluvial Episode might have improved from 232 ± 11.6 million years ago (Mya) to 232 ± 0.116 Mya. An error of 116,000 years may still sound ridiculously long – but for a geologist it's a miracle!

The Wrangellia basalts can be dated directly because they are igneous rocks. They were molten and they solidified, and crystals within them can thus be dated to give the time at which they solidified. The sedimentary rocks that document the episode are harder to date directly in this way. However, in the Dolomites in northern Italy, there are wonderful sequences of marine sediments that have been dated into small zones of less than 1 million years each. Interleaved between these are terrestrial sediments with footprints, and these document how dinosaurs were absent before the Carnian Pluvial Episode, but then present in abundance after the event. We used these amazing fossil sites to argue that the Carnian Pluvial Episode triggered the second phase in the origin of the dinosaurs, their great explosion 232 million years ago, in a 2018 paper, led by my former doctoral student Massimo Bernardi, now curator of geology at the Museum of Sciences in Trento in north Italy.

The cross-dating between marine and non-marine sediments in north Italy is confirmed by a remarkable, and independent, method called magnetostratigraphy. This method relies on the fact that the Earth's direction of magnetization has flipped from north to south, and back again, many dozens of times through Earth's history. Nobody can quite explain why north flips to south, and what happens during the flip. Yet the record is there in magnetic minerals in the rocks, and the alignment of the striped barber's poles of normal-reversed-normal-reversed can fix the relative ages of rocks of all kinds.

Earlier, I suggested that the diversification of dinosaurs occurred in three phases. We've looked at the first two – their origin about 245 million years ago in the maelstrom of recovery from the Permian–Triassic mass extinction, and their explosive diversification 232 million years ago following the Carnian Pluvial Episode. The third phase followed the end-Triassic mass extinction 201 million years ago, and we will explore this a little more in the next chapter.

How can we identify ancient climates?

In describing the origin of dinosaurs, I have freely talked about arid and monsoonal climates. How are these important conclusions reached? The basics go back to the beginnings of geology. Sedimentology is the science of understanding sediments and reconstructing ancient environments. First-year geology students learn to distinguish marine and non-marine rocks. Marine rocks, for example, uniquely contain minute fossils of plankton, and larger fossils of animals that only lived in the sea, such as brachiopods, sea urchins, or sea lilies. Rocks deposited in lakes or rivers may, on the other hand, contain leaves, insects, or dinosaurs. Of course, leaves, insects, and dinosaurs can be washed down rivers into the sea, but it's the predominant fossils, not the rarities, that count. The rocks of the Ischigualasto Formation, for example, include tree trunks and leaves from conifers and other plants, which confirms they were deposited on land. The sediments are red-coloured mudstones and sandstones. In places, there are large channels formed by ancient rivers, and muds deposited in temporary lakes. There are also burrows constructed by some of the smaller reptiles, and this confirms chemical evidence that the area experienced really hot conditions at times, which drove the little beasts underground for protection.

There are all sorts of other clues in the rocks. For example, sand dunes of certain types mean deserts. Channels and stacked sediments can identify meandering rivers, flipping from side to side through time. Layers of salt can indicate coastal pools drying under a hot sun.

There are also chemical indicators of ancient conditions. For example, isotopes of oxygen measured through whole rock sections can track rising and falling temperatures. The ratio of oxygen isotopes varies with temperature because of differential effects during evaporation from the surface of a pond or as rain falls, and the isotope signals can also reflect salinity and the volume of water locked up in ice sheets.

Environments of deposition can be identified locally, but what about the worldwide picture?

How different was the Triassic world from ours?

The Earth is composed of numerous great tectonic plates, which are constantly in motion. Some of the plates underlie the continents, and others make up the sea floor. The engine for plate movements comes from the molten mantle of the Earth, which lies below the solid plates. Great convection currents rotate within the magma, transferring their lateral motion to the solid crust. In places, molten material from the mantle comes to the surface, such as along the great mid-ocean ridges. Down the centre of the North and South Atlantic is a continuous system of fissures through which basalt lava bubbles from time to time. The mid-Atlantic ridge surfaces on Iceland. The constant supply of fresh crust in the centre of the Atlantic, and in similar ridge systems in the Pacific and Indian oceans, drives the ocean floor plates apart at the rate of about 1 centimetre (3/8 inch) per year. Where plates move past each other, there can be great faults, such as the San Andreas Fault through California that periodically judders into life. This is real-time evidence that the Earth's crust is in constant motion. In other places, oceanic plates dive beneath continental plates, such as along the Pacific coast of South America.

During the Triassic, all continental plates were fused together as the supercontinent Pangaea. Also, there was no land across the poles, so there

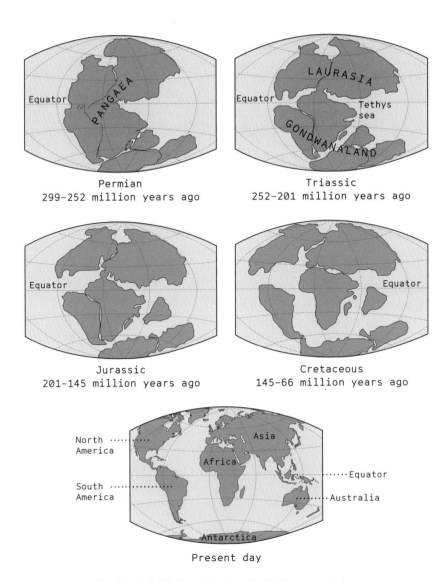

Permian
299–252 million years ago

Triassic
252–201 million years ago

Jurassic
201–145 million years ago

Cretaceous
145–66 million years ago

Present day

Continental drift from the Permian to the present day.

were no ice caps. This means that temperatures from equator to pole varied less than they do today, and the climate is often said to have been equable. Uniform climates over a single land mass meant that the early dinosaurs and other land animals and plants could be much more widely distributed then than now.

In the Late Triassic, as we have seen, there was a violent set of huge volcanic eruptions along the west coast of Canada, pouring out the magma that solidified as the Wrangellia basalts. Ten million years later, another set of similarly huge volcanic eruptions were set off in the middle of Pangaea, along the rift line of a new ocean that was jerking fitfully into being. These eruptions also produced great thicknesses of basalt, most famously seen today in the Palisades along the Hudson River between New York and New Jersey.

All down the coastal strip of eastern North America, there are rift valleys, extending from Nova Scotia in the north and east to North Carolina in the south. These rift valleys formed as the Earth's crust was splitting by the force of great convection cells in the mantle that were pulling what is now Europe and North Africa eastwards, and what is now North America westwards. Even though the rate of movement was only 1 centimetre per year, over thousands of years the tension became too great, and the crust ripped apart – just as is happening today in the Great Rift Valley of east Africa, as that continent tears apart too.

The Late Triassic rift valleys of eastern North America formed lakes, and great thicknesses of lake sediments were laid down, often containing fish, insect, and plant fossils and, every now and then, a trackway produced by a dinosaur as it hurried across the wet lakeside sediments.

Asia, Europe, and North America were in the northern hemisphere in the Late Triassic, but all somewhat south of where they are now. London and New York were at Mediterranean–Caribbean latitudes, and so they were considerably warmer than they are today. The absence of ice caps also ensured there were no cold winters. South America, Africa, India, Antarctica, and Australia were all joined together in the southern hemisphere, and these formed continuous land across the equator to the northern continents. Dinosaurs could march over thousands of kilometres from South Africa to Arizona, or from Canada to north Africa. There were regional faunas in some places, defined probably by mountain ranges and climate belts, but mostly the plants and animals on land were much more capable of spreading worldwide in the Triassic than today.

These conditions continued into the Jurassic. Even though the north Atlantic had begun to unzip, animals could walk across from Africa to South America, and there were still some connections from North America through Greenland to Europe. This was true right through to the Late Jurassic, 150 million years ago, when dinosaurs such as the huge sauropod *Brachiosaurus* is known from Tanzania in east Africa and Wyoming in the central United States. The huge predatory dinosaur *Allosaurus* is also known from Wyoming, and possibly from Tanzania, but also from Portugal. There was a wide ocean between the northern and southern continents along the line of the equator, but the traffic of dinosaurs seems to have been funnelled north and south through a little strip of land that linked Morocco in the south and Spain in the north.

During the Cretaceous, the continents continued rotating and moving. The south Atlantic opened, and the traffic of land life between South America and Africa was eventually cut off. The southern continents separated, with Africa moving north and keeping some contacts with Europe, but the southern tip of South America linking to Antarctica to the east, and then Australia. In the Late Cretaceous, India broke free and began its long trek north to eventually dock with the rest of the Asian continent about 50 million years ago. India continues to drive north, . forcing the Himalayas ever higher as it does so.

In the Late Cretaceous, not only were the continents moving closer to their present positions, but also sea levels rose hugely, by some 100 metres (nearly 330 feet) or more, as a result of enhanced mid-ocean mantle activity and uplift. This sea level rise flooded coastlines round all the continents, and split Africa and North America with mid-continental seaways. This meant that dinosaurs in the Late Cretaceous had their opportunities for migration massively curtailed. For example, the famous *Tyrannosaurus rex* of this time is found only in North America, not elsewhere in the world, unlike many of its predecessors. For the first time, east coast dinosaurs could not even cross the North American continent to meet their cousins on the west coast.

....

Changing climates and changing worlds. A few years ago, dinosaur palaeontologists thought that all the main outlines of the origin of the dinosaurs had been resolved. Then everything changed. New fossils pushed dinosaur origins back by 15 million years into the Early Triassic. Perhaps, if you had been transported back in a time machine, you would have barely noticed the first dinosaurs. Among all the abundant, hefty, and noisy rhynchosaurs, synapsids, and crurotarsans snuffling around, the few, small, bipedal dinosaurs nipping in and out of the undergrowth would have seemed like a sideshow.

Their explosion onto the scene 15 million years later, after the devastation wrought by the Carnian Pluvial Episode, was dramatic. Dinosaurs replaced rhynchosaurs and others overnight, geologically speaking. Understanding how all this happened has involved some remarkable new fossil finds, but also extraordinary advances in our understanding of rock dating and our ability to reconstruct ancient climates and ancient worlds, and to use modern computational methods to crunch the data and test the models of large-scale evolution.

This is a chapter that will definitely need rewriting in ten years' time. I predict someone somewhere will find some of these oldest dinosaurs, which are somewhat elusive right now. New studies in the field will pin down the nature of the Carnian Pluvial Episode better, and new analyses will help us properly understand the big evolutionary and ecological changes that were tearing the Earth and life apart through the Triassic.

Chapter 2

Making the Tree

Classifications cause controversy. Through my entire research career, palaeontologists have squabbled strenuously over the classification of their organisms of choice, whether it be dinosaurs, trilobites, or fossil plants. These fights might seem inconsequential, but we are considering the fundamentals of how to document the wonders of biodiversity, and we are also addressing origins.

Documenting biodiversity and origins is big science now – indeed, it forms part of the modern techniques termed, rather forbiddingly, phylogenomics and bioinformatics. Phylogenomics is the new discipline of establishing evolutionary trees from molecular data. Bioinformatics is the field of managing large data sets in the life sciences and number-crunching those data to produce information on the genetic basis of disease, adaptations, and cell function, and has applications fundamental to medicine and agriculture. Practitioners of these methods block their university's supercomputers for weeks while they run billions of repeat calculations to get their answers. The American National Science Foundation has invested millions of dollars in their 'Tree of Life' initiative – a programme to fund consortia of scientists to produce complete evolutionary trees of particular groups of plants and animals, such as all 11,000 species of birds or all 300,000 species of flowering plants.

We need accurate inventories of species to plan for practical conservation measures. Understanding which of the many species of mosquito passes on malaria parasites was critical in seeking cures. Biomedical scientists study evolutionary trees of fast-evolving viruses such as AIDS and influenza. In the case of viruses, evolutionary trees span months or years, whereas the evolutionary trees constructed by dinosaur palaeobiologists span millions of years. Without evolutionary trees, sometimes called phylogenies, we cannot explore big patterns in evolution. So while in some ways the classification of life might seem trivial or merely an arcane branch of librarianship, in others it is crucial.

In this chapter, we will follow the quest I shared with a number of other contemporaries to crack the family tree of dinosaurs. The story began in 1984, at a conference in Tübingen, southern Germany, where four of us turned up a little nervously with our independent first efforts. We were all then in temporary employment, having finished our doctoral theses a year or two before, and eking out livings variously as research fellows in the United States (Jacques Gauthier and Paul Sereno) and the United Kingdom (Dave Norman and myself). Were we on the right track or not? We had each tackled some aspect of dinosaur phylogeny and all agreed, independently, that dinosaurs had all evolved from a single ancestor, and between us we had a solid picture of the exact pattern of relationships of the main dinosaur groups. This was a first, but would these iconoclastic ideas be accepted?

Since 1984, we have continued our quest for more and more complete understanding of the dinosaur family tree. In 2002 and 2008, teams in my laboratory produced the first supertrees of dinosaurs – each the result of hugely laborious number-crunching – purporting to show the complete set of relationships among hundreds of species. And, unexpectedly, the whole thing blew up again in 2017 when a radical new proposal tore apart the consensus on dinosaur relationships. This is a story of discovery of new fossil specimens, innovative new ideas about methods, the harnessing of improving computer power, and the continuing fascination of the dinosaurian tree of life. It's not over yet.

Back in 1984, the risk of rejection for our first tentative dinosaur family trees was heightened by the fact that we had all applied a revolutionary new set of techniques called cladistics. We coded our data on primitive punch cards and sent them off to our university's mainframe computer services, and then waited a few days for the results to come in. This was all massively controversial in the 1980s, and we have since lived through decades of refinement of methods. We need to look back to the beginning of the so-called 'cladistic revolution', and consider whether the risks we took paid off. First, though – why was it all so controversial?

What was the cladistic revolution?

The cladistic revolution was about methods. When I was trained in classification, we used textbooks from the 1960s by luminaries such as Ernst Mayr and G. G. Simpson. They agreed that the best approach to classifying species, whether living or extinct, was to apply a great deal of experience. As Mayr recounted, it took him decades as a researcher on bird biodiversity to learn that a character such as feather colour would not help much in determining the deep relationships of birds, but that more fundamental characters, such as the shape of the bill or some particular muscles of the wing, were much more useful. As Simpson said, 'classification of species is more art than science'.

While Mayr and Simpson were writing, the cladistic revolution had already begun, but neither they nor anybody else noticed. In the year 1950, Willi Hennig, a rather dour professor of entomology in Berlin, published a book he had been writing in the 1940s while a prisoner of war. He did not use the term 'cladistics'; that was introduced about 1960 by other evolutionary biologists, from the Greek word *klados*, which means a branch, referring to the 'branches' of an evolutionary tree. Hennig, in fact, called his new discipline 'phylogenetic systematics', meaning that he wanted to explain that the process of reconstructing the tree of life should be more science than art – he wanted biologists and palaeontologists to delve into their data, and discriminate the true worth of the characters or traits that they wished to use for the fundamentals of classification.

Hennig's book was in German, but very few biologists, whether they could read German or not, paid much attention. It was only when it was translated into English in 1966 that his message was received. At first, the evangelists for the new method, researchers at the American Museum of Natural History in New York and Natural History Museum in London, were intensely excited and wrote eloquently on the topic. They also applied the new methods to their studies of mutual interest, namely the evolution of fishes. However, others struggled, because Hennig had a very dry prose style, and he invented a lot of new terminology, often compound words, for his new ideas; readers had difficulties with the text both in German and in English. Nonetheless, Ernst Mayr and G. G. Simpson both waded in with highly critical attacks on the new cladistics, and Mayr invented the term 'cladist', which he and others used to designate (and denigrate) the adherents of this new creed.

The museum evangelists in England and the United States pushed the ideas and sought to explain them through publications and conferences. By 1983, when I was writing my doctoral thesis, it was far from clear that Hennig's cladistics would prevail. Most biologists and palaeontologists were indifferent or hostile to the new idea. I remember attending a meeting of the Willi Hennig Society in London in 1984, just before the Tübingen meeting, where people were shouting at each other, and one speaker was twirling the microphone on its cord and threatening the chairperson, who was trying to shut him up. On other occasions, tempers became so frayed that public apologies had to be demanded and delivered.

Why so much heat and so little light? Willi Hennig's insight was rather straightforward: that we need a testable method for the construction of phylogenetic trees, and this should be based around phylogenetically informative characters. Palaeontologists were to stop searching for ancestors, because you can never test a hypothesis of ancestry, but instead to seek out sister groups: that is, nearest relatives.

As we saw in Chapter 1, the Silesauridae family is sister to the Dinosauria. This is a big claim: that these two groups are closest relatives and that they share a close common ancestor. In a modern representation, we can show the relationships of the dinosaurs to silesaurids, and all their other close relatives among the archosaurs, as an explicit tree, or cladogram (see overleaf). Groups fit within groups, and each group has a single ancestor and is characterized by one or more phylogenetically informative characters. These unique groups with single points of ancestry are called clades: hence the terms cladistics, cladogram, and cladist. In the case of the hypothesis that silesaurids and dinosaurs are closest relatives, the evidence is that they both share six or seven unique anatomical characters not seen in any other animals, such as the proportions of their hip bones, an opening between the ischium and pubis in the hip area, modifications to the femur and tibia in the hindlimb, and a rising process on the front of the astragalus bone in the ankle. We don't consider general characters such as their slender limbs, their pointed teeth, or other features that are seen widely among early reptiles – these are like Mayr's feather colours, not informative in constructing the cladogram.

The search for the phylogenetically informative character is a tough one, but it provides a focus for testing. Anyone who wishes to argue that Silesauridae is not the sister group of Dinosauria must present an alternative hypothesis, in the form of a different cladogram that is supported by better evidence comprising alternative anatomical characters. In general, the more informative anatomical characters that can be mustered, the more likely the hypothesis is to be correct.

This is what many of the critics did not like. They were being forced to go further than they were comfortable in doing. It was no longer good enough to hide behind the fig leaf of a dashed line and a scattering of question marks. The dinosaur family trees in my university textbooks showed all the relationships of the main dinosaur groups confidently in the Jurassic and Cretaceous, but the lines then tailed off into a cloud of uncertainty as we got to their origins. How far back should you go?

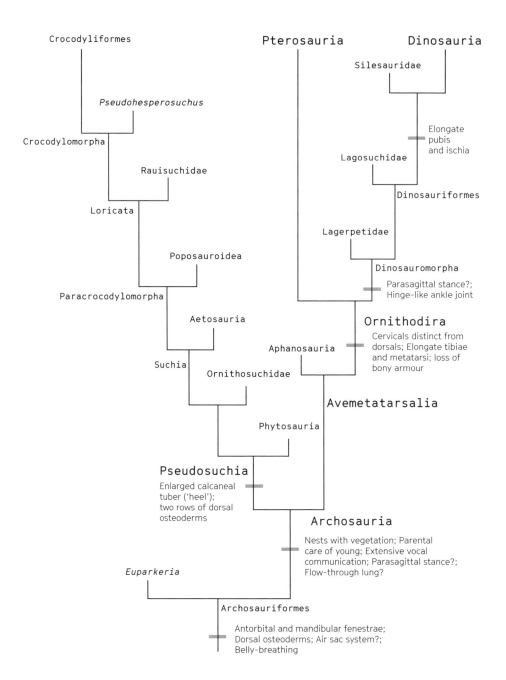

Crocodyliformes

Pseudohesperosuchus

Crocodylomorpha

Rauisuchidae

Loricata

Poposauroidea

Paracrocodylomorpha

Aetosauria

Suchia

Ornithosuchidae

Pterosauria

Dinosauria

Silesauridae

Elongate
pubis
and ischia

Lagosuchidae

Dinosauriformes

Lagerpetidae

Dinosauromorpha

Parasagittal stance?;
Hinge-like ankle joint

Ornithodira

Cervicals distinct from
dorsals; Elongate tibiae
and metatarsi; loss of
bony armour

Aphanosauria

Avemetatarsalia

Phytosauria

Pseudosuchia

Enlarged calcaneal
tuber ('heel');
two rows of dorsal
osteoderms

Archosauria

Nests with vegetation; Parental
care of young; Extensive vocal
communication; Parasagittal stance?;
Flow-through lung?

Euparkeria

Archosauriformes

Antorbital and mandibular fenestrae;
Dorsal osteoderms; Air sac system?;
Belly-breathing

Cladogram showing the evolution of archosaurs, with key
phylogenetic characters indicated. The very close relationship
between Dinosauria and Silesauridae is clear.

Discovery of the clade Dinosauria

Understanding about dinosaur classification had gone back and forth over the years. Richard Owen first named the group Dinosauria in 1842, and he included the theropod *Megalosaurus*, the sauropodomorph *Cetiosaurus*, and the ornithischian *Iguanodon* – one of each of the fundamental subgroups. Then in 1887, Harry Seeley, a professor at the University of Cambridge, dropped a bombshell. He had studied all the dinosaurs known to Owen, plus many more that had been found since 1842, and had decided they did not form a natural group. Rather, he claimed they should be split into two groups, the Saurischia, for the theropods and sauropodomorphs, and the Ornithischia. He noted that the saurischians all shared the so-called 'reptile hip' and the ornithischians all shared what he called the 'bird hip'.

Seeley's insights were partly right, partly wrong, but it was the wrong aspects that dominated the subject for nearly 100 years. The reviews and textbooks showed dinosaurs as having arisen from two, three, or even more distinct ancestors. This showed a lack of clarity of thinking. Of the two styles of dinosaurian hip, only that of ornithischians was unique to the group. The saurischian hip arrangement was seen in crocodiles, lizards, and, perhaps confusingly, also in birds. In cladistic terms, it's

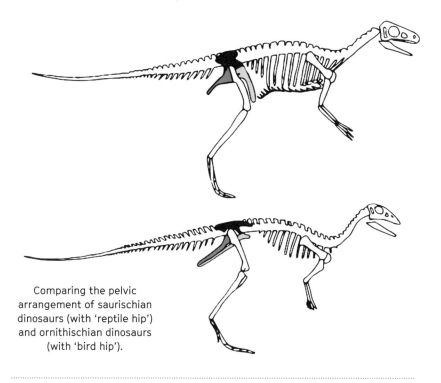

Comparing the pelvic arrangement of saurischian dinosaurs (with 'reptile hip') and ornithischian dinosaurs (with 'bird hip').

impossible to demonstrate that the saurischian hip arrangement was not simply acquired from their ancestors without modification. By the way, the so-called 'bird hip' of Seeley was a misnomer – although a unique character complex of Ornithischia, it was independent from the hips of birds, which in fact evolved from theropod dinosaurs, as we shall see later.

The confusion was partly resolved in 1974, when the American palaeontologist Bob Bakker and the English palaeontologist Peter Galton argued for the validity of Owen's Dinosauria. In their opening words:

> Traditionally dinosaurs are classified as two or three separate, independent groups of reptiles in the Subclass Archosauria. But evidence from bone histology, locomotor dynamics, and predator/prey ratios strongly suggests that dinosaurs were endotherms [= warm-blooded animals] with high aerobic exercise metabolism, physiologically much more like birds and cursorial mammals than any living reptiles.

However, their conclusion was expressed in terms of shared aspects of biology and ecology, and that was not enough to convince the doubters. After all, the shark and the dolphin share many features of their swimming mode and feeding style, but that doesn't change the fact that one is a fish and the other a mammal.

This was the background all four of us had before the 1984 Tübingen meeting: eighty years of rejection of the reality of dinosaurs as a natural group by all the experts, and a cheeky proposal against that view by Bakker and Galton. We were convinced Bakker and Galton were right, but what was needed was a properly worked-out cladistic hypothesis, with each branch in the tree supported by one or more bullet-proof anatomical characters. In my paper, I identified fourteen characters unique to Dinosauria, including a series of features of the hindlimb, such as the inturned head of the femur, the muscle-bearing processes of the femur, the roller-like astragalus in the ankle with a rising process in front of the tibia, the much reduced calcaneum in the ankle, the bunched toe bones, and posture up on tippy-toes. These all relate to the fact that dinosaurs had perfected a fully upright stance at the time of their origin. The fact that saurischians and ornithischians share all these detailed anatomical features was convincing evidence, we thought back in 1984, that all dinosaurs formed a single natural group, Dinosauria, with a single ancestor.

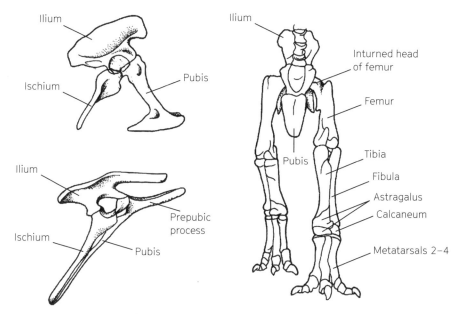

Ilium

Ischium

Pubis

Ilium

Ischium

Pubis

Prepubic process

Ilium

Inturned head of femur

Femur

Pubis

Tibia

Fibula

Astragalus

Calcaneum

Metatarsals 2–4

The key characteristics of dinosaurs
are in the hindlimbs.

For the Tübingen meeting, we had each come up with a similar list
of unique features of Dinosauria. I stopped at the origin of dinosaurs, but
Dave Norman and Paul Sereno went further, and proposed outline trees
for Ornithischia, whereas Jacques Gauthier did the same for Saurischia.
Nervous as we were – and we did receive some criticism at the meeting –
it seems the time had come. Bakker and Galton had taken a great deal of
the flak in 1974, and so ten years later, when we presented such knock-
down evidence – coming from four of us independently – the world
was perhaps ready to listen. We all published more detailed papers later,
documenting all the evidence, and this became the textbook norm.

In detail, it turned out I had been wrong about many of the
supposedly unique dinosaur characters – in fact, many of them were
present also in silesaurids and other close relatives of dinosaurs. My
only excuse is that many of these beasts had not yet been discovered
in 1984, and when they were later presented to the world, they caused
a redistribution of characters up and down the cladogram; but it did not
affect the overall hypothesis.

Having established the tree of dinosaurs (see overleaf), we should
introduce some of the key players, following the narrative through
geological time.

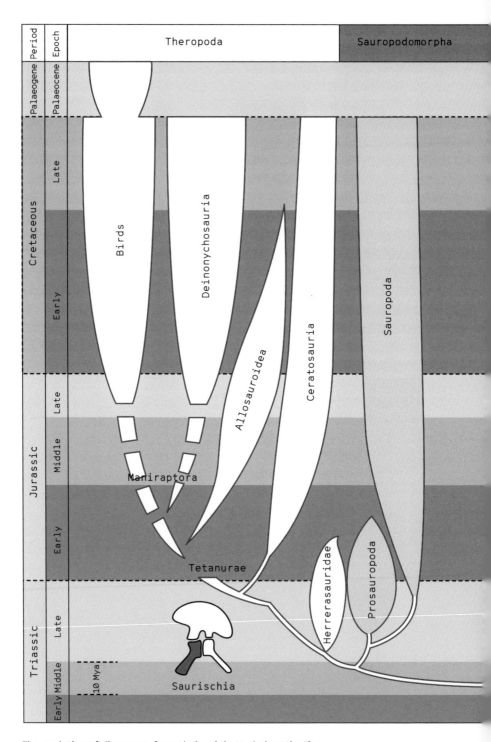

The evolution of dinosaurs, from their origin to their extinction.

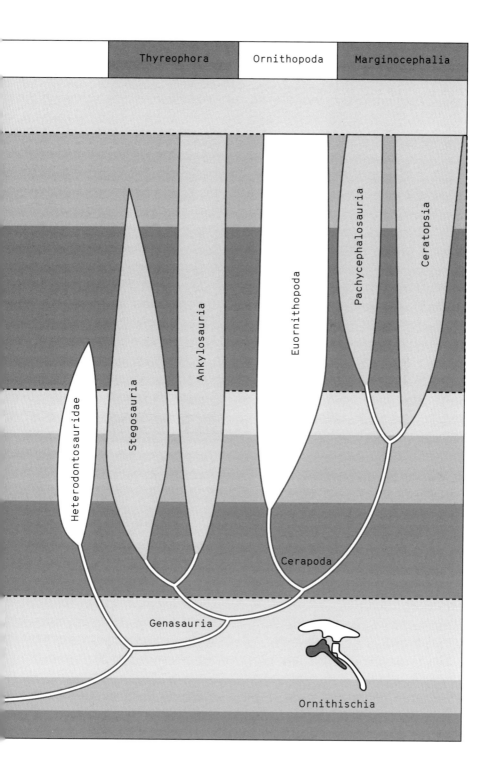

Thyreophora

Ornithopoda

Marginocephalia

Heterodontosauridae

Stegosauria

Ankylosauria

Euornithopoda

Pachycephalosauria

Ceratopsia

Cerapoda

Genasauria

Ornithischia

The Triassic explosion

As we saw in Chapter 1, dinosaurs originated deep in the Triassic –
the period from 252 to 201 million years ago – and then diversified in
two or three steps. By the time of the Late Triassic, many key forms had
emerged. This is clearly demonstrated in the Trossingen Formation in
southern Germany, dated at about 215 million years ago. Fossils have been
collected from this 40-metre-thick (130-foot) unit of yellow-coloured
sandstones, most notably in some large excavations undertaken in the
1920s near Stuttgart.

Picture the Late Triassic in southern Germany. The landscape is quite
flat, with abundant wet-climate plants such as horsetails, ferns, and seed
ferns growing around the rivers and lakes, and dry-climate conifers
around the hills. A two-legged theropod, *Liliensternus*, flits past, chasing
a small lizard. *Liliensternus* is 5 metres (16½ feet) long, slenderly built, and
with a long, narrow skull. It snaps at the lizard, but its prey darts away.
Then, a great thundering is heard, as a pack of much larger dinosaurs
bursts onto the scene. *Liliensternus* crouches among the plants, looking
out for juveniles it could pick off.

Genus:	*Plateosaurus*
Species:	*engelhardti*

The new dinosaurs on the scene are a herd of **Plateosaurus**, some twenty individuals ranging in size from babies, barely a metre long, to juveniles 5 metres (16½ feet) long, and hoary old adults 10 metres (33 feet) in length. The *Plateosaurus* mainly stand up on their hind legs, but flop back down onto all fours to feed on horsetails by the river bank. The feet have four toes that spread wide to support the body weight. The hands also have four main fingers, and the thumb claw is large and flattened. With its broad curve, the *Plateosaurus* rake up plant food on the ground, before stooping to snatch it in their jaws. Their skull is long, almost horse-shaped, with a nostril at the front, and a long snout, with the jaws lined with twenty-five strong, leaf-shaped teeth above and below. They snip leaves with their front teeth, and tip their head back to push the leaves down into their throats and swallow the chopped plant fragments. When they stand up to look around, they throw their head and neck back, drop their tail towards the ground, and hoist their heavy front quarters off the ground.

A tree falls, and the herd of twenty shoot away at speed. They mostly lift their hands from the ground, and with neck stretched forward, tail stretched back, and the entire backbone horizontal, they charge off, one of them nearly kicking the *Liliensternus* aside.

Named by:	Hermann von Meyer, 1837
Age:	Late Triassic, 227–210 million years ago
Fossil location:	Germany
Classification:	Dinosauria: Saurischia: Sauropodomorpha: Plateosauridae
Length:	up to 10 m (33 ft)
Weight:	1 tonne (2,205 lbs)
Little-known fact:	Herds of *Plateosaurus* skeletons were once thought to have died in arid desert, but they were actually trapped in soft mud.

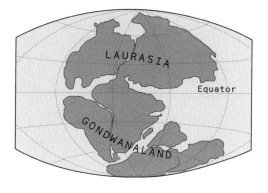

This was the world of the Late Triassic, and it did not last. Even though, in parts of the world, sauropodomorphs such as *Plateosaurus* were abundant, making up large herds, they and the others died out at the end of the Triassic, 201 million years ago. Great volcanic eruptions began along the sides of rifts opening up between what is now Europe and Africa on the one side and North America on the other. In the Triassic, as we saw, all continents were fused as one, Pangaea, but at the end of the Triassic, the Atlantic Ocean began to unzip. This process was driven by great eruptions of basalt lava along straight-line rifts, the precursor of the mid-ocean ridge of today, which we can observe onshore as it slashes across Iceland.

The basalt lavas belched out continuously over thousands of years, and with them came gases such as sulphur dioxide and carbon dioxide, which mixed with water in the atmosphere to produce acid rain. This killed plants on land, and the landscape was swept clear of forests and soils. The oceans became acidified, and this killed animals with carbonate shells. The volcanoes also poured out other gases, including methane and water vapour, which, with the carbon dioxide, led to sharp greenhouse warming, and this in turn drove life from the tropics and also removed oxygen from the ocean floor. This is the model of one of the big mass extinctions, the so-called 'end-Triassic event', which saw the end of many dinosaur groups, but also many of the other tetrapods (four-limbed animals) that, as we saw in Chapter 1, had been part of the landscape. Life recovered, and the extinction marks a major boundary: the end of the Triassic and the beginning of the Jurassic.

The Jurassic world

There was a lot going on in the Jurassic, the time from 201 to 145 million years ago. In the Triassic, as we saw, the three main lines of dinosaurs had become established: the theropods, sauropodomorphs, and ornithischians. All three groups branched out substantially in the Jurassic. The meat-eating theropods, all active predators, and mostly with sharp teeth lining their jaws, diverged into some small forms that headed into the trees, evolved feathered wings, and became flyers, some of which we now call birds. Other theropods evolved to be larger and larger, adapting to the hunting opportunities offered by the larger herbivores.

Among these plant-eaters were the long-necked sauropodomorphs, and these included some Late Jurassic giants, weighing up to 50 tonnes,

I Late Triassic scene in Arizona, showing four examples of the small flesh-eating dinosaur *Coelophysis*, and a couple of plant-eating sauropodomorphs behind, being menaced by a rauisuchian (a large archosaur).

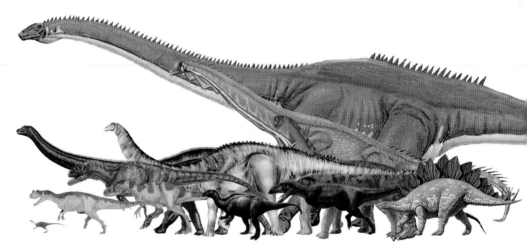

II March-past of dinosaurs from a single time, the Morrison Formation (Late Jurassic) of Wyoming, showing the range in size, from tiny *Ornitholestes* at the front to mid-sized *Stegosaurus* with plates, and giant *Diplodocus* at the back.

III Dinosaurs in a strangely modern landscape. Flowering plants began to take over most terrestrial landscapes in the Late Cretaceous, and these dinosaurs of the Hell Creek Formation of Montana enjoyed the sights and smells of magnolias and roses – but they still largely fed on the ferns and conifers they had known for millions of years.

IV Replica of the famous London
specimen of *Archaeopteryx* showing
the skeleton and feathers on the
wings and tail.

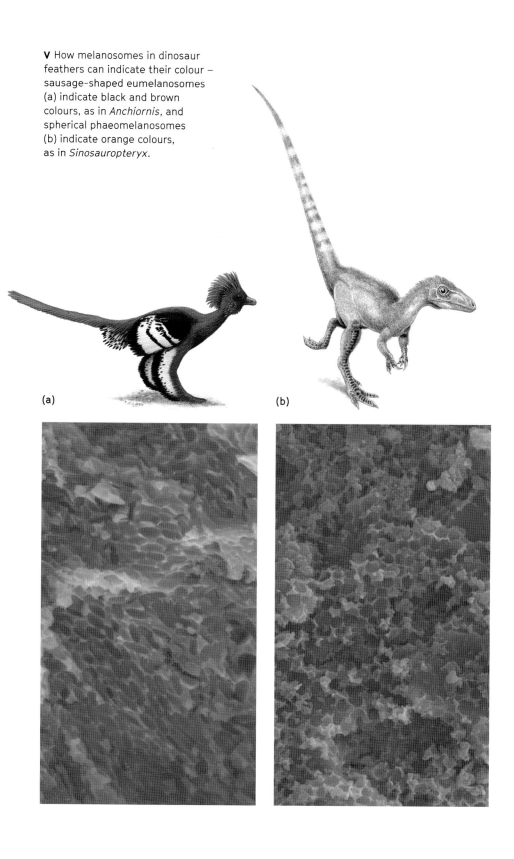

▼ How melanosomes in dinosaur feathers can indicate their colour – sausage-shaped eumelanosomes (a) indicate black and brown colours, as in *Anchiornis*, and spherical phaeomelanosomes (b) indicate orange colours, as in *Sinosauropteryx*.

(a)

(b)

VI A dinosaur tail in amber, with all the bones and dried-up muscles inside, and a rich covering of feathers. The fossil (below) shows an ant and other debris caught in the amber, and the feathers show every detail of barbs and barbules (above).

VII Reconstruction of the little theropod whose tail got caught in amber 125 million years ago. Like many other small theropods of the time, it was heavily covered in feathers, and chased bugs on the ground and in the trees.

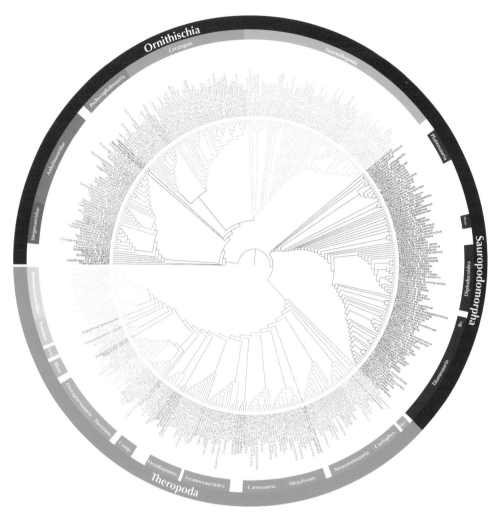

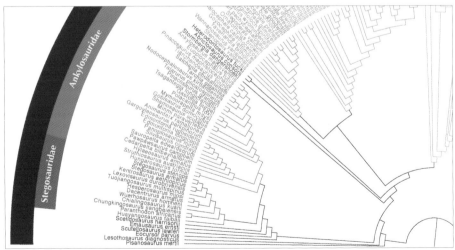

VIII Supertree of all the dinosaurs – an evolutionary tree showing dinosaur origins at the centre, and the expansion of the group into ornithischians (red), sauropodomorphs (blue), and theropods (green).

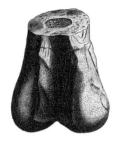

The first dinosaur bone ever illustrated, from Robert Plot's 1677 *The Natural History of Oxford-shire*.

and clearly so large that they were probably not preyed on by any other dinosaur. The third group, the ornithischians, a group of herbivores, some bipedal and others quadrupedal and armoured, were not so diverse in the Jurassic, but two armoured groups did emerge during this time: the stegosaurs, with plates and spines down their backs; and the ankylosaurs, encased in a rather heavily armoured cuirass over the whole body.

I first encountered Jurassic dinosaurs when I was a Junior Research Fellow at the University of Oxford. We were taken on field trips to the local Middle Jurassic by the excellent Curator of Geology at the University's Museum of Natural History, Phil Powell. Phil was a man of many talents, renowned for his expertise on the bagpipes. He would practise on his chanter (the tubular part of the bagpipes with fingering holes) at lunchtime in the museum. As a Scot, I appreciate the bagpipe, but it is really an outdoor instrument, best played at some distance from the audience.

Phil Powell took us to the limestone quarries around Oxford from which numerous members of the university had recovered dinosaur bones. Indeed, the first dinosaur fossil ever recorded came from a small quarry in Cromwell parish, north of Oxford, but it was not at first recognized for what it was. The specimen was the lower end of a femur of the theropod **Megalosaurus** (see overleaf), showing two bulbous facets, and broken across to reveal the internal structure. It was illustrated by Robert Plot, Keeper of the Ashmolean Museum and Professor of Chemistry at Oxford University in his classic *The Natural History of Oxford-shire*, published in 1677. Here, he illustrated many genuine fossils, as well as curiously shaped stones, some like horse heads, or human kidneys and feet. He identified the dinosaur bone as a leg bone from a very huge human being.[1]

---

1 This first-found dinosaur bone was later named *Scrotum humanum*, the first formal Latin name ever given to a dinosaur, by Richard Brookes, in 1763. Sadly, this was to become a *nomen oblitum* ('forgotten name'), and the dinosaur was later given the monicker *Megalosaurus bucklandii* in 1824. But for the fact that the name *Scrotum humanum* was not widely used, and so became forgotten, the Megalosauridae – the family of dinosaurs containing *Megalosaurus bucklandii* – might instead have been named the Scrotidae.

Genus:	**_Megalosaurus_**
Species:	_bucklandii_

We have never been able to collect a complete specimen of _Megalosaurus_, but the various skeletal parts found are enough to show that it would have measured about 9 metres (30 feet) long and might have weighed 1.4 tonnes (3,086 pounds). This was one of the first truly large predatory theropod dinosaurs, capable of preying on nearly all the other dinosaurs of its day. It habitually ran on its muscular hind legs, the great three-toed feet spreading wide as it paced along, and leaving extensive trackways that can still be seen in the mud of quarries around Oxford. The arms were powerful, and presumably used in grappling with prey.

The fauna of the Middle Jurassic of central England included the theropod _Megalosaurus_, the sauropod _Cetiosaurus_, and the armoured, plant-eating ankylosaur _Lexovisaurus_. These were accompanied by a rich

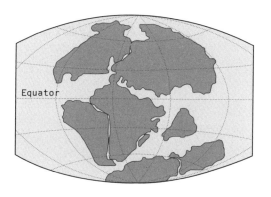

Equator

Named by:	**William Buckland, 1824 (genus);** **Gideon Mantell, 1827 (species)**
Age:	**Middle Jurassic, 174–164 million years ago**
Fossil location:	**England**
Classification:	**Dinosauria: Saurischia: Theropoda:** **Megalosauridae**
Length:	**9 m (30 ft)**
Weight:	**1.4 tonnes (3,086 lbs)**
Little-known fact:	**The first fossil of *Megalosaurus* was** **reported in 1676, and named *Scrotum*** **humanum in 1763.**

fauna of smaller animals, including salamanders, lizards, crocodiles, pterosaurs (the flying reptiles of the time), and early mammals, which have been remarkably well documented in ancient mines around the village of Stonesfield, 24 kilometres (15 miles) northwest of Oxford. The Stonesfield mines had been worked for centuries for roofing slates, and the miners followed the appropriate layers deep underground. They hauled out the rough limestone slabs, set them up to freeze and split, and then sold them as a safer, but weightier, roofing material than thatch. Since the 1820s, Stonesfield had yielded numerous fossils, including dinosaur teeth, as well as bones and teeth of the small critters from the age of dinosaurs, attracting attention from palaeontologists very early in the history of the discipline.

New discoveries of Jurassic dinosaurs from China

Dinosaurs are known throughout the Jurassic, with rich finds of Early Jurassic forms from North America, South America, and South Africa. The Middle Jurassic was only patchily known – largely from England – and I was intrigued to follow up the Middle Jurassic in China, where astonishing finds had been made since the 1990s.

My chance came in 2016, when I was invited by Professor Baoyu Jiang from the University of Nanjing to accompany him on a field trip to Inner Mongolia, in the north of China. We were to visit various sites in the Tiaojishan Formation, which extends over much of the south of Inner Mongolia, and neighbouring Hebei and Liaoning provinces. Our special focus was a unit within this formation, the Daohugou Bed, the source of an amazing assemblage of plants and animals. I was excitedly anticipating finding dinosaurs and pterosaurs, but in fact we spent several weeks splitting rocks and finding insects. These were impressive enough, however, including hand-sized cockroaches, beetles, flies, and water boatmen. Jiang had employed teams of local farmers to help, and we squatted in the burning heat under tarpaulin awnings, splitting rocks. I chipped away, and inevitably did not find much, while one of the local Mongol farmers amassed a huge pile of excellent specimens. Professor Bo Wang, the palaeoentomologist from the Nanjing Institute of Geology and Palaeontology, was excited, and he could see opportunities for new work. I was fretting for dinosaurs.

Up to 2016, eleven dinosaurs had been reported from the Daohugou Bed, as well as pterosaurs, lizards, and other reptiles. The dinosaurs were all astonishing, being mainly small tree-dwellers, most of them with feathers, and equipped for gliding in one way or another. One, which rejoices in the shortest name ever given to a dinosaur, is *Yi qi*, a real oddball, with all the normal limbs of a small dinosaur, and feathers, but also special struts along the arm that strongly suggest it had bat-like membranes as well to aid gliding and capture of insects.

The most famous Daohugou dinosaur is **Anchiornis** (see overleaf), now known from dozens of skeletons, and one of the first dinosaurs to have had its colour determined (Chapter 4), reconstructed as a proud turkey-like animal, with long, black tail, legs sporting feathered banners behind like some crazy cowboy trousers, wings with long black and white striped feathers, and a ginger tuft of feathers on its head. I wanted to find one.

I failed.

Each day, my haul of insects was respectable, but far short of those being unearthed by the farmers and students. Dr Jiang was keen to acquire specimens of dinosaurs and pterosaurs, so we arranged to meet various dealers. On several occasions, I was brought along to comment or evaluate (which I could not really do), and we would meet a dealer, often quite expert in the local fossils, and acting as an intermediary for the farmers. Slabs were brought out of the boots of cars, or we visited workshops in anonymous tower blocks to see the wares. This is the way palaeontology works in China, because the hours that scientists and their students can spend in the field are limited, and the local farmers have the time and the eye to make amazing discoveries.

We were shown some wonderful *Anchiornis* specimens, but Dr Jiang wanted something really special, something new. In the end, in an unlit room in an industrial town in neighbouring Hebei Province, we were shown a beautiful pterosaur specimen, which had been cleaned up but not coated in glue or preservative. This is important when you want to run the specimen under the scanning electron microscope to identify tiny structures or run chemical analyses. This specimen proved to be important in determining something about the evolution of feathers in dinosaurs and their relatives, as we shall see.

The dinosaur world becomes busier and more familiar in the Late Jurassic, with the classic Morrison Formation dinosaurs of the Midwestern United States, including all the favourite sauropods such as *Diplodocus*, *Brontosaurus*, and *Brachiosaurus*, theropods like *Allosaurus* and *Ceratosaurus*, the plate-backed *Stegosaurus*, and many more (see pl. ii). The Morrison dinosaurs had first been excavated in the 1870s, when railroad crews were cutting through the mountains and plains of Wyoming, Colorado, and Utah. Great crates full of these dinosaur bones were sent back to the east coast, where they can still be seen in the museums of Philadelphia, New Haven, New York, and Washington. The density of fossil finds at some localities tells us these dinosaurs were abundant. It's nearly impossible to imagine the giants of the Late Jurassic world. The largest sauropod, *Brachiosaurus*, would walk over you without noticing – its belly was 2.5 metres (8 feet) above the ground, and it towered to a height of 9 metres (30 feet), reaching around in the tree tops for leafy branches to eat. The other sauropods of the Morrison (such as *Diplodocus*, with their long, horizontally held necks and whip-like tails, and *Brontosaurus* and *Camarasaurus*) were nearly as large, but more heavily built.

When a mixed herd of these giant sauropods got moving, the thunderous noise and rising dust would have been incredible. The flesh-

Genus:	**Anchiornis**
Species:	*huxleyi*

Named by:	Xu Xing and colleagues, 2009
Age:	Middle Jurassic, 166–164 million years ago
Fossil location:	China
Classification:	Dinosauria: Saurischia: Theropoda: Maniraptora: Anchiornithidae
Length:	40 cm (16 in.)
Weight:	0.7 kg (1½ lbs)
Little-known fact:	*Anchiornis* had small wings on its hind legs, which would have looked like cowboy trousers when it walked along.

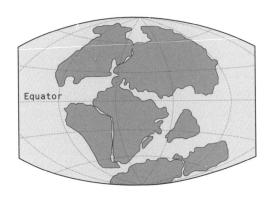

Equator

eating theropods *Allosaurus* and *Ceratosaurus* were both large enough, at 8.5 metres (28 feet) long, that smaller dinosaurs would have fled before them. The giant sauropods were vulnerable only when young; once they had reached an age of five years or more they probably would have been immune to predation. Similar Late Jurassic faunas are known also from Tanzania, Portugal, and China, representing a time when sauropods in particular were distributed worldwide and reached their largest sizes. This was not to last, however.

Foodwebs and the Cretaceous heyday of the dinosaurs

There were major changes on the Earth about 145 million years ago, and the rock successions in certain parts of the world show evidence of shifts of continental plates and climate change. The sauropods, which had dominated Late Jurassic faunas, dropped in diversity and the ornithopod dinosaurs became the dominant herbivores. These include the classic ornithopod **Iguanodon** (see overleaf), named in 1825 – the second dinosaur ever named – from Sussex in southern England.

Iguanodon is one of the most-studied dinosaurs – well, probably not quite as much as *T. rex*! Nonetheless, dozens of specimens have been found, mainly in Europe, and many publications have been made. Most notably, Dave Norman, one of the four of us who were struggling with the first dinosaur cladograms in 1984, has made *Iguanodon* his life's work. The specimens are beautiful, and many skeletons are complete. *Iguanodon* was typically 10 metres (33 feet) long and weighed about 3 tonnes (6,615 pounds). It was built as a biped, but commonly went down on all fours to feed. The hands had five strong fingers for grasping, but they bore small hooves – perfect evidence that they were sometimes used in walking. The spiky thumb was probably used in defence. The massive hind legs made *Iguanodon* a powerful bipedal runner, with the tail stretched out straight behind. In fact, the balancing function of the tail is enhanced by thin rods of bone that run down on either side of the backbone, stopping the tail from waving about too much at speed.

Iguanodon had an elongate skull, with a deep snout, and it snipped leaves with the toothless bony plates at the front of its jaws. It passed the fragments back to be roughly chopped between the long, straight tooth rows down each side of the jaw. It was the immense capacity of *Iguanodon* and its relatives to process plant food that made them so successful. This was the first dinosaur ever to be able to *chew* its food. Other dinosaurs

Genus:	*Iguanodon*
Species:	*bernissartensis*

had simply grabbed and swallowed, but by chewing its food, *Iguanodon* could extract much more goodness from every mouthful. It did not chew the food as we do, by rotating the lower jaw around the pivot at the back, but more by a mechanism that allowed the lower jaw to chop up into the upper jaws, a little like the blade of a penknife shutting into the handle. As the lower jaw cut upwards, the teeth ground across the upper set of teeth, both sets sharpening each other as they did so. Further, the cheeks could expand a bit outwards, so providing some lateral grinding.

Since the first find in the 1820s, *Iguanodon* has been found at numerous localities throughout the Weald area of southeast England, and the Early Cretaceous rocks of this area were soon called Wealden, and were identified widely across Europe. The Wealden was a time of warm, damp climates, and the rocks record life on the lowlands, with rich remains of lush plants, insects, amphibians, lizards, crocodiles, dinosaurs, and even some birds and mammals.

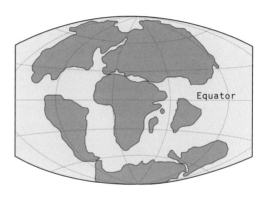

Equator

Named by:	Gideon Mantell, 1825 (genus); Louis Dollo, 1881 (species)
Age:	Early Cretaceous, 140–125 million years ago
Fossil location:	England, Belgium
Classification:	Dinosauria: Ornithischia: Ornithopoda: Iguanodontidae
Length:	10 m (33 ft)
Weight:	3 tonnes (6,615 lbs)
Little-known fact:	The most complete collections of *Iguanodon* came from the roof of a coal mine in Belgium.

The Wealden sediments accumulated up to 700 metres (2,300 feet) thick, as sands and muds washed down from highland areas around what are now London and Belgium. As the Weald basin sank, the sediments piled up, and they span 15 million years of the Early Cretaceous, from 140 to 125 million years ago. The sediments record ancient rivers in some cases, deltas and shallow marine incursions in others. After years of detailed study, Percy Allen from Reading University was able to make a detailed palaeoenvironmental reconstruction of the Wealden scene, showing lakes, rivers, crevasse splays, trapped logs, and charting how the sediments and fossils accumulated.

The Wealden fossils give us a much clearer picture of Early Cretaceous life. Fossils have been collected from the Weald for over 200 years, and these include logs and tree trunks found at some levels, complete skeletons of *Iguanodon* and other dinosaurs elsewhere, and bonebeds of microvertebrates. The microvertebrates – tiny bones and teeth, as the

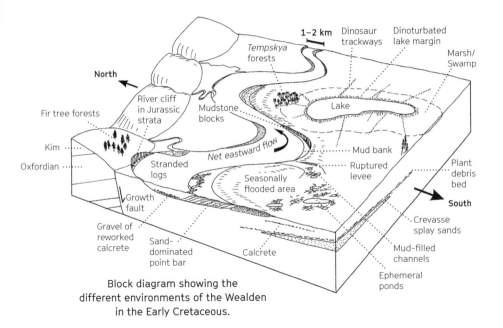

Block diagram showing the different environments of the Wealden in the Early Cretaceous.

name suggests – may seem less spectacular than the huge bones, but they provide a rich record of diversity at the time. If you truly want to understand ancient faunas and floras, you must include everything.

One researcher who sampled the Wealden microvertebrates, Steve Sweetman, discovered a remarkable range of creatures including sharks and bony fishes, salamanders, frogs, the most diverse lizard fauna yet discovered from the Early Cretaceous, turtles, crocodiles, pterosaurs, ornithischian and saurischian dinosaurs, birds, and mammals. The tiny teeth may look rather bewildering, even if startlingly beautiful, but they all show diagnostic characters that, after a while, palaeontologists learn to recognize, and which show in some detail how many species were present, and what they were all up to.

How to put this all together? The food web is a classic way to represent all the species in an assemblage, and especially who ate whom. Steve Sweetman has kindly shared his lifelong knowledge of the Wealden ecosystem by preparing the food web shown opposite. This kind of diagram makes the past come back to life, and the Wealden ecosystem was similar to those today in some ways, profoundly different in others. The main difference, of course, is the role of diverse dinosaurs, and the absence of birds and large mammals.

The Wealden shows us a very typical picture of life in the times of the dinosaurs, but here and there were hints of changes to come. Among the ferns, seed ferns, conifers, and other vegetation were some unusual incomers – the world's first flowers.

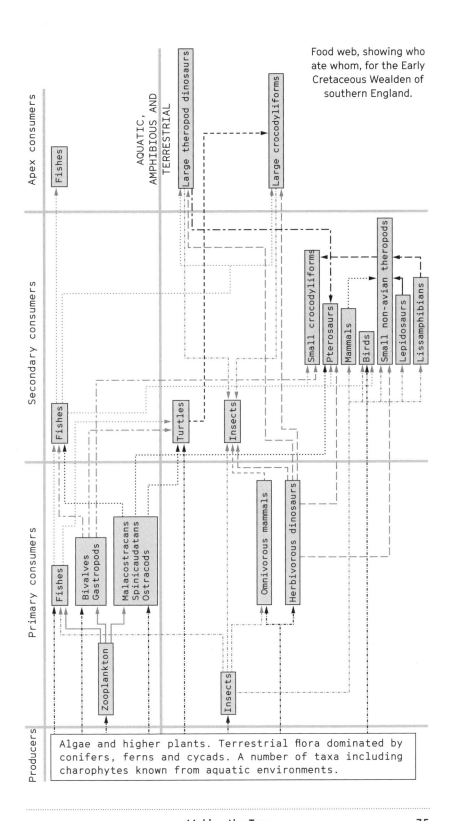

Food web, showing who ate whom, for the Early Cretaceous Wealden of southern England.

Apex consumers

Fishes

AQUATIC, AMPHIBIOUS AND TERRESTRIAL

Large theropod dinosaurs

Large crocodyliforms

Secondary consumers

Fishes

Turtles

Insects

Small crocodyliforms

Pterosaurs

Mammals

Birds

Small non-avian theropods

Lepidosaurs

Lissamphibians

Primary consumers

Fishes

Bivalves
Gastropods

Malacostracans
Spinicaudatans
Ostracods

Zooplankton

Omnivorous mammals

Herbivorous dinosaurs

Insects

Producers

Algae and higher plants. Terrestrial flora dominated by conifers, ferns and cycads. A number of taxa including charophytes known from aquatic environments.

Constructing the dinosaur supertree

This is the place to come back to evolutionary trees. We have reviewed some of the early stages in the history of dinosaurs, and the effort to establish the broad outlines of dinosaurian evolution. By the year 2000, some 500 species of dinosaurs had been named from the Triassic, Jurassic, and Cretaceous of nearly all countries of the world. They could all be assigned quite easily to theropods, sauropodomorphs, or ornithischians, but what about the finer-scale patterns of relationships? When we thought about this question, around 2000, biologists were already well advanced in drawing up ever more detailed trees of their favourite groups.

We felt it was important to construct a complete tree of all dinosaur species using as much information as we could. This had not been attempted before. In this case, rather than looking at all the dinosaur specimens, coding the characters of their skeletons and skulls, we thought we would try a new set of computational procedures and construct a so-called supertree. A supertree, as the name suggests, is built from many regular trees. So, my then PhD student, Davide Pisani, now a professor at Bristol, took the lead. We scanned every paper written between 1980 and 2000 about the relationships of dinosaurs, and we collected together 150 such trees.

The theory of building a supertree is quite simple, although in practice there are many headaches in achieving an agreed result. If you have two evolutionary trees, each of ten dinosaurs, and two of the species are shared in both trees, you fix the trees together using these common species, and make a tree of eighteen species. Of course, among the 150 trees, there were lots of proposed relationships that disagreed, so the computational exercise Davide set in train had to crunch through these disagreements and try to identify the most likely solution. After weeks and weeks of computing, we had come up with a complete tree of 277 dinosaurs. We couldn't include all 500 known dinosaur species because, at that time, many had not yet been included in any kind of cladistic analysis.

The tree was pretty well resolved, meaning that most dinosaurs were in a specific position, but there were quite a few places where a bunch of five or six species all branched off from a single spot. These were places where disagreements existed, and we could not make the program give us a more precise answer. It was frustrating, but realistic – often there just is not enough information to decide definitively one way or the other. These places where relationships are unresolved are a useful marker for researchers to go hunting for more information.

Six years later, we did it again. This time, the project was led by Graeme Lloyd, another Bristol doctoral student who is now on the staff at the University of Leeds. Graeme was able to identify 550 papers with dinosaur trees of one kind or another, and including 420 species. The number-crunching was even more brutal, but we were very proud of our much-improved dinosaur supertree, now in full colour, and drawn as an attractive circle (see pl. viii). When we submitted the work for publication, we were crestfallen that the journal declined to print our lovely diagram – they said it was too big to fit on the page.[2] Nonetheless, with 420 species, this was one of the largest supertrees available at the time. Now, it is commonplace to produce massive trees – the biggest perhaps comprising all 11,000 species of birds.

Is this just a child's game – my supertree is bigger than yours? Well, yes, in a way. There is the technical side and the desire to get it right, and to push the software and computing hardware to the limits. Significantly, though, we could use the supertree for studies of macroevolution. In particular, we asked a simple question, which was, when were dinosaurs evolving fast? Graeme Lloyd looked at every one of the 423 branching points in the whole tree and calculated whether it showed an unusual rate of evolution, in terms of the number of species that had branched from that point over a known amount of time.

He identified only 11 of the 423 branching points as showing evolutionary rates that were statistically faster than expected. Seven occurred in the Late Triassic, and another two in the Middle Jurassic and two in the Late Cretaceous. This bottom-heavy aspect of the dinosaur tree was unexpected – it shows that, in a broad sense, dinosaurs had done much of their evolving in the first half of their history, and not so much after that.

We took this a step further, and compared dinosaur evolution in the Cretaceous with the evolution of other groups of land-living plants and animals. The slow rate of dinosaur evolution in the Cretaceous stood out as unusual – something pretty major was going on at that time, and we named it the Cretaceous Terrestrial Revolution. Dinosaurs, unexpectedly, were not part of this dramatic evolutionary step.

2   We have made the tree available online, and you can see it in all its glory at http://zoom.it/JJLR – this website allows you to move the tree around, and zoom in to see details of each species.

The Cretaceous Terrestrial Revolution: the trigger for modern life

About 125 million years ago, flowers revolutionized the Earth; indeed, they triggered what has come to be known as the Cretaceous Terrestrial Revolution, the time when terrestrial ecosystems, like that of the Wealden, were remodelled radically. The key questions are to determine the scope and impact of the ecological revolution, and to determine the extent to which the huge changes in plants and animals did, or did not, affect dinosaur evolution.

The Cretaceous Terrestrial Revolution marked the point in the entire history of life when life on land first became hugely diverse. It's been worked out that in the Early Cretaceous there were about equal numbers of species on land and in the sea, whereas now life on land is five to ten times as diverse as life in the sea. Most of this huge diversity of modern life is made up from insects, but other very rich groups include spiders, lizards, birds, and the flowering plants themselves.

There may be as many as a million species of beetles on Earth. The great twentieth-century English biologist J. B. S. Haldane was once asked what he had learned about Creation from his long studies of nature. He replied that evidently 'God has an inordinate fondness for beetles'. Indeed. We don't even know how many beetle species there are – some 400,000 have been named, and whenever a beetle expert is let loose in a new bit of jungle, he or she comes back with fifty new species each day. There is a limit to how fast any individual can work and publish the descriptions and names...so it will take a few centuries to get on top of this pile of unfinished work.

Anyway, in total, there may be 15 million species of all kinds on Earth, and a good 80–90 per cent of these are on land, with only 10–20 per cent in the sea.

When biologists produce evolutionary trees of modern organisms, most of these highly species-rich groups (flowering plants, beetles, butterflies, bees, bugs, spiders, lizards, mammals) seem to have radiated explosively in the mid-Cretaceous, about 100 million years ago. What was going on?

The driver was the explosion in numbers of flowering plants, or angiosperms as they are properly called. Angiosperms include nearly all familiar plants, from fruits to vegetables, oaks to palms, and all the grasses. They are economically crucial to humans as virtually all grains and pulses we eat are angiosperms. It is said that the angiosperms diversified in the Cretaceous because of their unique fertilization system, which involves flowers, seeds contained within nutritious fruits. This fertilization system gave them immediate advantages over other plants, and they were able to adapt to new settings and survive crises better than, for example, the conifers or ferns.

From the first, angiosperms engaged in mutual relationships with pollinators such as birds and bees...but also butterflies, moths, and wasps. Bugs and other insects adapted to feed on the succulent new leaves, flowers, and stems. Thus, some 300,000 species of angiosperms today support 2 million or more species of rather specialist insects, and the angiosperms make lush, complex forests, much, much richer in species than comparable conifer-dominated forests, which today tend to occur in colder climates.

Did the Cretaceous Terrestrial Revolution affect the dinosaurs? The new food sources could have given them opportunities to adapt and diversify. However, the consensus is that they were not much affected. The dinosaurs stomped around as ever, probably spurning the flowering plants, trampling over their perfumed flowers to get a good mouthful of crunchy ferns or spiky conifer leaves.

There were new dinosaur groups at the time, but they did not seem to depend on flowering plants, or any of the new insect groups. As we move from the Early to the Late Cretaceous, the iguanodons and other dinosaurs of the Wealden were replaced by hugely abundant duck-billed hadrosaur dinosaurs, some such as **Parasaurolophus** (see overleaf) with remarkable head crests. The function of these crests has been debated. The crests were hollow, made from the bones of the snout, so they could not have been for fighting or defence. The best explanation is that they were for species recognition – at any time, a hadrosaur herd might include five or six species, each characterized by its own headgear – meaning that any individual would want to associate in herds with members of its own species. Like birds today, the dinosaurs were probably visual animals, and so used the same kinds of cues to identify species as we do.

Equally common in some places were the ceratopsians, the horned-faced dinosaurs, like huge rhinos. Other plant-eaters included the tank-like, armoured ankylosaurs, some with great whacking clubs on their tails, as well as a few long-necked sauropods, especially in southern continents. The predatory theropods included a huge array, from tiny feathered insect-catchers, through the long-limbed ostrich-like ornithomimosaurs, to the greatest Jaws of the lot, *Tyrannosaurus rex*, apex killer of the latest Cretaceous of North America.

Genus:	*Parasaurolophus*
Species:	*walkeri*

After this survey of dinosaurs through time, everything seemed to be more or less settled. Palaeontologists believed they had wrestled most of the dinosaurs into their correct positions in the great tree of life. Then came a bombshell in March 2017 – it seemed that the consensus on fundamental dinosaurian relationships had been completely wrong.

Dinosaur trees and the evolution revolution

The bombshell was a paper written by Matt Baron and Dave Norman at the University of Cambridge and Paul Barrett at the Natural History Museum in London that really set the theropod among the pigeons.

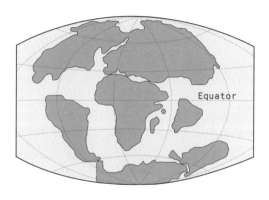

Equator

Named by:	**William Parks, 1922**
Age:	**Late Cretaceous, 76–73 million years ago**
Fossil location:	**United States, Canada**
Classification:	**Dinosauria: Ornithischia: Ornithopoda: Hadrosauridae**
Length:	**9.5 m (31 ft)**
Weight:	**5.1 tonnes (11,244 lbs)**
Little-known fact:	**The crest used to be misinterpreted by some as a breathing snorkel, but the top end is sealed.**

In this paper, the authors claimed that palaeontologists had been wrong about the deep classification of dinosaurs, and presented a radical new dinosaur tree. A few months later, a flurry of papers re-ran tests of their new hypothesis, some agreeing with it wholeheartedly, and others (including one I co-authored) casting some doubts.

The Baron paper attracted huge attention. It was the subject of a front-cover painting in *Nature*, and was reported worldwide, with headlines such as 'After 130 years, the dinosaur family tree gets dramatically redrawn' in the *Atlantic* and 'Shaking up the Dinosaur Family Tree' in *The New York Times*. The *Guardian* reported, in classic *Guardian* style, that this was 'A discussion, not a war: two opposing experts talk dinosaur family trees'. The subject rumbles on, without resolution yet.

The standard view, since 1984 at least, was that Dinosauria consists of two subgroups, Saurischia, comprising theropods plus sauropodomorphs, and Ornithischia. In their new paper, Baron and colleagues argued that the three main dinosaur groups were arranged differently, with Theropoda flipping to pair with Ornithischia, rather than Sauropoda. The new clade of Theropoda + Ornithischia was called Ornithoscelida, and the Sauropoda were left outside. In our riposte, led by Max Langer from the University of São Paulo, we checked through the huge data matrix assembled by Baron and colleagues and picked holes here and there, and then re-ran the analysis, and recovered the traditional arrangement of Ornithischia and Saurischia – but only just.

A word about how cladistics is done. In studies such as these, the input of effort is not trivial. Matt Baron included seventy-four species in his analysis – not every dinosaur ever named, of course, but more than enough to provide broad coverage of all clades, as well as some silesaurids and other relatives. Each of these seventy-four species was coded for 457 anatomical characters, meaning that Matt (or someone else) had visited numerous museums around the world, pulled out the drawers, and checked each character, usually for a yes/no answer – character present or absent. These are usually coded as '1' for present and '0' for absent. Therefore, Matt Baron ended up with a huge data sheet with 74 rows and 457 columns, for a total of 33,818 cells that had all been checked. In our revision, we ploughed through some of these, focusing on specimens we could access easily, and so were able to correct some of the coding.

What is the point of these huge data compilations? They are, in fact, the basis for the calculation of the best-fitting tree. There are many ways to number-crunch through big data sets like this, seeking the tree, or more likely set of trees, that most efficiently, or most probably, accounts

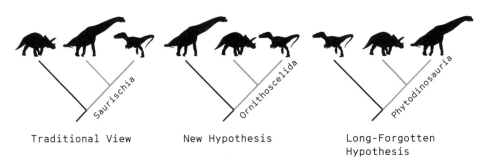

Traditional View   New Hypothesis   Long-Forgotten Hypothesis

The three possible arrangements of major dinosaurian clades.

for most of the information. With so much data, it is not likely that a single phylogenetic tree will emerge, nor that it will fit the data perfectly. Much more commonly, there may be 100 or more equally most likely trees and these have to be summarized.

So, in the end, the to-and-fro of papers in 2017, and later in 2018, commenting on the radical new Baron proposal, did not lead to a clear answer. What was agreed was that more work should be done – close scrutiny is needed of all those 457 anatomical characters to make sure they are all independent and phylogenetically informative, and then teams of experts need to go back to the museum drawers and check over the fossils in detail.

Even then, there may not be a clear result. This might sound shocking, or an indictment of the cladistic method. It is certainly no criticism of cladistics, though, because the alternative to cladistics (or an equivalent statistical approach) is...nothing. We'd be back to assertion and guesswork – 'I think the ankle tells us the story', 'No, I think the skull characters are more important'.

What we may be seeing here is a star phylogeny, as it is sometimes called, a series of branching points in the phylogenetic tree that perhaps happened very fast, or for which we lack critical fossils. A star phylogeny is an explosion of diversity, evidence of fast evolution of a new clade, and in some cases, perhaps including this one, there was no time for any unique anatomical characters to arise; or they might have been overwritten by later evolution. Perhaps it will be forever difficult to identify the key features that demonstrate once and for all whether the nearest relatives of theropods are sauropodomorphs or ornithischians.

....

Evolutionary trees, or phylogenies, are the key to understanding evolution. The details of how trees are constructed may seem arcane, and indeed in the past fifty years mathematicians and computer scientists have contributed massively to improvements in the function and speed of the methods. However, the transition from evolutionary trees drawn by hand, and really little more than informed guesswork, to computer-generated cladistic trees has been one of the most profound examples of how dinosaurology has shifted from speculation to science.

The reader may find the to-and-fro of the debates and trees, whether cladistic trees or supertrees, heavy going, but the consequences are far-reaching. These trees are the essential underpinning of how we describe dinosaur evolution through the Triassic, Jurassic, and Cretaceous, and – importantly – how we make calculations of relative rates of change. Identifying the fact that dinosaurs did most of their evolving in the first half of their time on Earth, and then slowed down, is profound. It may be wrong, of course, but the counter-argument can only be made by identifying faults in the original analysis and providing a better hypothesis using better methods and better data.

It's extraordinary, after our initial efforts back in 1984, and the long decades of reworking and stabilization of the dinosaur tree, that it has all been blown apart again. It will certainly take the combined efforts of numerous experts to check through the data and explore possible solutions between the Saurischia and Ornithoscelida models of fundamental dinosaur relationships.

Chapter 3

Digging Up Dinosaurs

I got excited about dinosaurs when most kids do, about the age of seven or eight. This enthusiasm has never gone away, and it is strongest when I climb into a bouncing four-wheel-drive, in some hot and exotic corner of the world, and we head out into the field. The thrill of planning, reading around, spreading out the maps, and deciding where to go, cannot be beaten. It is a privilege, too, not only to work with professional colleagues in so many countries and continents, but also to live among the local people, and to be there not as a tourist, but as a person on a mission. Most exciting of all is the thrill of wondering what you might find.

Fieldwork is a standard part of any university degree in geology or biology, and I had spent plenty of time squelching around Scotland, following loping professors, and looking at obscure bits of grey rock buried beneath damp fronds. Visiting Elgin, 105 kilometres (65 miles) northwest of Aberdeen, where I grew up, was more promising. Here, the rocks at least were yellow, almost honey-coloured in the watery sun.

Footprints of an early reptile from Clashach in northeast Scotland.

More importantly, they had produced skeletons of ancient reptiles, and in the coastal sandstone quarry at Clashach, you could still see some footprints on the rocks. There, reptiles large and small had tramped up the lee slopes of the ancient sand dunes, presumably looking for water and plant food, and their prints had survived for over 250 million years, as fresh in every detail as the day on which they had been made. But northeast Scotland was not Mongolia, or Australia, or Canada.

My chance came when I was an undergraduate at Aberdeen University. I rather cheekily attended the conference of the Society of Vertebrate Palaeontology and Comparative Anatomy in 1976, held that year at University College, London. I was a mere undergraduate, but thought, why not? The professors attending the conference were kind, and did speak to the few gawky students such as myself who were there. During one of the tea breaks, I buttonholed a quiet American professor. He was J. Alan Holman – the 'J.' was just an initial – and, quite amazingly, he invited me to go into the field with him from his base at Michigan State University. This was my first trip abroad, at the age of twenty-one, and I spent July to September of the summer of 1977 in Michigan and Nebraska. Holman was then the leading expert on fossil snakes and lizards of North America, and he did a two-month field season each year working through fossil beds in the Valentine Formation. He employed me as his field assistant, and even paid me to dig tonnes of sediment and dump it into great sieves constructed in wooden crates, which we agitated in the river. This washed away the mud and left rocks, twigs, and fossils behind. We boxed up the concentrate and took it back east for sorting and classifying. The humid heat of Nebraska was a shock to a pallid Scot, and the minuscule fossils weren't quite dinosaurs, but this was living.

After returning to the United Kingdom, I wrote to Phil Currie, then a young researcher at the University of Montreal, who had just got his first job at the University of Alberta in Edmonton. He was the dinosaur man, and is now arguably the greatest living dinosaur expert in North America, or at least one of the top two or three. Currie responded by similarly offering me a job, paid, as his field assistant for the summer of 1978, and we lived for two months in the remote desert-like parts of southern Alberta around Drumheller. This was in Dinosaur Provincial Park, which had been established in 1955, but before the Royal Tyrrell Museum of Palaeontology had been established (it opened in 1985). I have since worked in the field in Germany, Romania, Russia, Tunisia, and China, but the principles of finding and digging up dinosaurs are the same everywhere.

How do palaeontologists find dinosaurs?

The key to finding dinosaurs is to choose the right kind of rocks – they must be the right age and it helps if dinosaurs have been found there before. Dinosaur Provincial Park in Alberta was a good choice, as many skeletons had been excavated there over the previous century. Once you are in the right kind of territory, the secret is good prospecting.

We drove the 280 kilometres (174 miles) from Edmonton to Drumheller in our field truck, a white pick-up with room for three in the front and a flat bed at the back to carry a few tonnes of bones. In the cavalcade was also a sleeping trailer, with beds for six, and a basic kitchen where one of the staff prepared exceptionally salty food and soups. When we complained, he told us we needed salt to replace the electrolytes we were losing in the heat of the sun; you don't argue with the cook.

The first task I learned was prospecting, walking up and down the coulées. These are deep ravines that have been washed into the landscape by the occasional heavy rains to which this part of Alberta is subject. They cut down through soil and sandstones. The rocks belong to the Dinosaur Park Formation – what else could it be called? Rock formations are units of sedimentary rock (usually) that have a definite bottom and top, stratigraphically speaking, and can be mapped.

The Dinosaur Park Formation is a unit about 70 metres (230 feet) thick comprising green-grey sandstones and mudstones deposited in the latest Cretaceous, some 75 million years ago, in terrestrial environments. The sediments have yielded leaves and trunks of trees, river-dwelling molluscs and fishes, as well as dinosaurs of course – some forty species of them, including the horn-faced ceratopsians *Chasmosaurus*, **Centrosaurus** (see overleaf), and *Styracosaurus*, the duck-billed hadrosaurs *Gryposaurus*, *Lambeosaurus*, and *Parasaurolophus*, the ankylosaur **Euoplocephalus** (see overleaf) with its tail club, the small, fast-moving predators **Ornithomimus** (see p. 90) and *Dromaeosaurus*, and the huge *Gorgosaurus*, 9 metres (30 feet) long, a close relative of *T. rex.*

The secret about prospecting for dinosaurs in the badlands is to look for scraps of bone, and follow them back upstream. The coulées had been washed out by erosion, and this happens repeatedly, so any trail of bone fragments in the bottom of a stream can be traced back up the branching streamlets to their source. Then, the job of the trip leader is to decide whether the prospect is worth excavating. Do we have a whole skeleton or just a fragment? You might have spotted just the final scraps of a skeleton, and nothing much would be left behind, or it could be the tip

| Genus: | **Centrosaurus** |
| Species: | *apertus* |

| Genus: | **Euoplocephalus** |
| Species: | *tutus* |

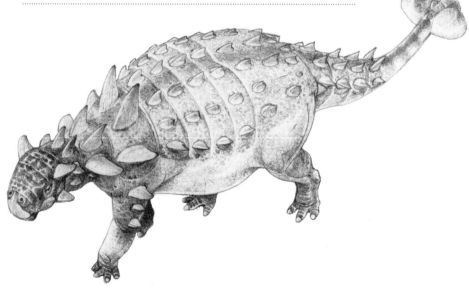

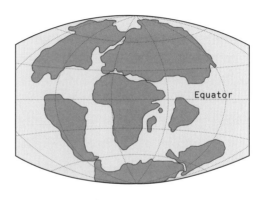

Named by:	**Lawrence Lambe, 1904**
Age:	**Late Cretaceous, 77–75 million years ago**
Fossil location:	**Canada**
Classification:	**Dinosauria: Ornithischia: Ceratopsia: Ceratopsidae**
Length:	**6 m (20 ft)**
Weight:	**2.5 tonnes (4,420 lbs)**
Little-known fact:	**One location yielding huge numbers of *Centrosaurus*, near Hilda, Alberta, is probably the richest dinosaur bone bed in the world.**

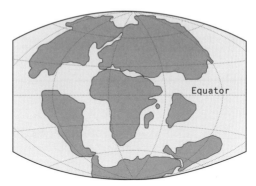

Named by:	**Lawrence Lambe, 1902 (species), 1910 (genus)**
Age:	**Late Cretaceous, 77–67 million years ago**
Fossil location:	**United States, Canada**
Classification:	**Dinosauria: Ornithischia: Thyreophora: Ankylosauridae**
Length:	**5.5 m (18 ft)**
Weight:	**2.3 tonnes (5,071 lbs)**
Little-known fact:	**This dinosaur was so heavily armoured that it even had bony eyelids to protect its eyes.**

Genus:	*Ornithomimus*
Species:	*velox*

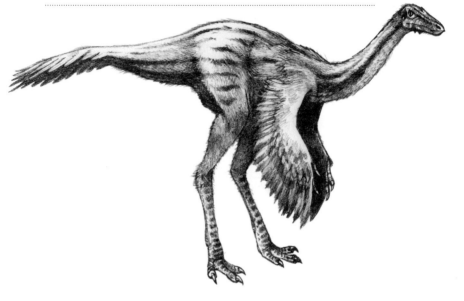

Named by:	Othniel Marsh, 1890
Age:	Late Cretaceous, 75–70 million years ago
Fossil location:	United States, Canada
Classification:	Dinosauria: Saurischia: Theropoda: Ornithomimidae
Length:	3.8 m (12½ ft)
Weight:	170 kg (370 lbs)
Little-known fact:	Ornithomimids did not have teeth so, although they are theropods, they may have had a mixed diet of small animals and plants.

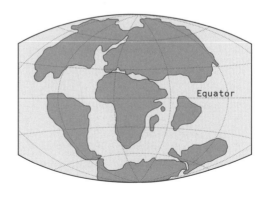

of the tail or a toe bone, and the rest is just waiting for you there, pristine, in the rock – unseen for 75 million years.

When the first dinosaur bones were found in the American West about 1860, the excavators were not trained scientists. In fact, they were navvies, driving the railroads through the open plains and mountains, and paid by the distance they could advance in a week. They were skilled at shifting rock fast. So, a whack with a great sledgehammer, or leverage by a long-handled spade, and out the bones would pop. They were thrown on flat-bed waggons and hauled by horse to the nearest railhead and sent east to the museums in New Haven, Philadelphia, and New York. Now, these no-nonsense methods would be frowned upon.

Several hours later, dripping with sweat in the roasting temperatures, we had all identified likely prospects. Phil Currie came round to inspect. One I had identified was chosen for our first excavation. The projecting bones showed it was probably a hadrosaur, one of the duck-billed plant-eaters that were hugely common in the Late Cretaceous, but worth extracting as the skeleton seemed complete, and suitable for exhibition.

The bones were lined up along a rather steep slope, so the first job was to build a bench in the rock, by levering and hacking the rock out from above the layer that contained the skeleton. You use any means at hand – we even had a huge and uncontrollable pneumatic drill operated by its own engine. It took a week to smash down the overburden to

Digging with power tools to remove rock from above the dinosaur specimen.

create a bench above the skeleton that would allow us to work with finer tools. We used hammer and chisel and power drills to remove the fine sandstone from above the bones. As we reached the fossils, we had to slow down and be careful, but it's hard to avoid any slip-ups, and on occasion the chisel gouged a chunk out of the bone – argh!

How do we record the excavation?

The first priority at excavation is to clear the whole site, so that the skeleton is laid out for view. Once the overburden is removed sufficiently, the site can be properly assessed. We could see the backbone of the dinosaur laid out, including the tail, limbs, and ribs. The skull wasn't there, however, and the neck was heading right into the cliff. So, we had to cover the site with tarpaulins, and push the cliff back further to extend the bench. For every foot of bench we cleared, we had to remove another yard height of cliff, the slope was so steep. Eventually, we had cleared the site back far enough (or at least we had compromised between shifting another 20 tonnes of overburden to retrieve one more bone and the risk of missing anything else buried under the cliff). At the end of a long day of digging in 30-degree temperatures, we welcomed the chance to rush down and jump into the Sandy River and soak in the cool waters.

Close-up of the hadrosaur, showing the tail flipped round along the backbone.

Gridding and mapping the hadrosaur skeleton.

Mapping came next. In those days, we did this by means of field sketches and photographs. We marked out metre squares across the site using strings, and these were the guide for close photography and drawing on squared paper. We could identify most of the bones – indeed the skeleton was pretty complete, and individual bones had only been rotated or moved a short distance by river currents.

Palaeontologists still, of course, draw and photograph their sites, but now they also commonly employ digital photographic techniques that record a perfect 2D or 3D model of the site, often called photogrammetry. The simplest form of photogrammetry is to take numerous overlapping photographs and then use standard software to blend the images together into a single large image covering the whole site. This software is like the landscape feature in many digital cameras that allows you to take a few photographs in sequence and merges them together.

More useful is 3D photogrammetry, where the photographs are combined to reveal the whole landscape, showing bones above and below each other, at different levels, and with sufficient fidelity to allow accurate measurements to be taken. The best efforts are achieved when a surveying set-up is used, with the camera on a tripod at fixed locations that are keyed to each other, so the angled photographs reveal all dimensions.

Photogrammetry is now commonly used to record dinosaur footprint sites, for example, where the intricacies of the depth and detail of each print can reveal something about how the dinosaur distributed its weight as it walked or ran (as we shall see in Chapter 8).

After the bones had been mapped, we began to remove them from the rock. This required planning and some risk. We could not simply lift the whole skeleton, as it would have been too wide across to fit on the truck, and in any case would have weighed more than 20 tonnes (44,100 pounds) in the rock. We were stuck up a steep-sided coulée, and no heavy lifting machinery could come close. There were some gaps between blocks of the skeleton, so we could dig deep cuts between these blocks, and we then created a system of deep trenches into the rock so that each cluster of bones was isolated on an upstanding island.

The next step is just what the early bone hunters soon learned. If you lever dinosaur bones out of the rock they break; covering them with a plaster cast would preserve them until they could be transported home. We mixed wet plaster in washing-up bowls, ran strips of burlap (sack cloth) through the plaster, and laid this criss-cross over the bones after

Use of power tools to remove rock from above the bones.

Applying sackcloth strips and plaster to strengthen the bone-bearing block.

protecting the bone surface with layers of kitchen paper. The idea was to build up a solid cocoon six or seven layers thick, and strengthen it with handfuls of plaster sculpted over the surface. After a day of plastering, your hands are dry and cracking.

The plaster cocoon would take several hours to set, and it had to lap down to the rock *below* the bones. Once primed, we inserted chisels and pry bars under the bone, and attempted to flip it. This is easy if the block is no more than human-sized, but to free the largest block we had to rig up a block and tackle above it, and run several chains into tunnels burrowed beneath. The block eventually flipped onto its side without fragmenting. Then, we cleared out all the rock we could from underneath the bones, and plastered the underside to make a complete, solid parcel.

The smaller bone parcels were carried by hand down to the truck. The medium-sized ones were mounted onto a curious contrivance consisting of a single bicycle wheel beneath a wooden frame, which could be guided by two people, one in front and one behind. This was precarious, and sometimes ran away, but nothing was damaged. The final block, weighing more than a tonne, sat in the debris of the excavation site, several miles from the nearest road, and 100 metres (328 feet) above the nearest point on the river bank to which the truck could approach. We therefore had to dig out a straight road down to the truck so it could line up with the

Clearing out loose rock from beneath
a 1-tonne (2,205 pound) flipped block
containing dinosaur bones.

Our pathway down to the truck; every step
of the 1980 dig was filmed.

dig site, and hauled the block down the slope using the vehicle's front-mounted winch and some long chains. We built a loading platform from rock, and managed to winch and coax the great block onto the load bed. If only we could have used a team of horses to drag the blocks on those steep slopes!

How are the bones extracted from the rock?

Back in the laboratory, the bone packages are laid out on benching, ready to be extracted. The plaster casing is removed using a small circular saw, and exposed bone is consolidated, usually with a soluble glue that soaks deep into the cavities, but which can be removed by solution in acetone (for example). A decent dinosaur preparation laboratory is laid out with numerous well-lit work stations, one for each technician. Each station is equipped with dental drills and other mechanical tools to remove rock, and there should be vacuum systems above the benches to remove the dust safely.

The physical removal of rock is done by sweeping a dental drill parallel to the bone surface – not aiming straight at it – in the hopes that chunks of rock will jump free, and thereby avoid nicking the bone surface with the drill. The technician keeps cleaning and consolidating the surfaces using soluble glue. You might think a fossil bone would be tough – well, they are, but tough and brittle, and consolidation is needed constantly to preserve the bone.

Each bone might take a day or more to free from the rock, and must be carefully numbered, tracked, and matched to the field map so they can be accurately reassembled later, if need be. The large block we collected took weeks to clean up, as many bones lay over each other, and each rib and vertebra had to be freed to get to the bones beneath. Sometimes, if the bones are too intimately entwined, they are left in the rock.

These methods are all long established, because they require the human eye and hand to work in coordination, there is no way to automate the process. However, technology now offers some amazing new opportunities. In the case of delicate structures, such as braincases, or small skeletons, the whole specimen can be X-ray scanned. This is computerized tomographic scanning, often shortened to CT scanning, in which the scanner captures X-ray images of the internal structure of the bone or rock, and these can be viewed as if they are a stack of slices, spaced maybe fractions of a millimetre apart. This means that museum

Back in the lab, clearing more rock from the plaster jacket.

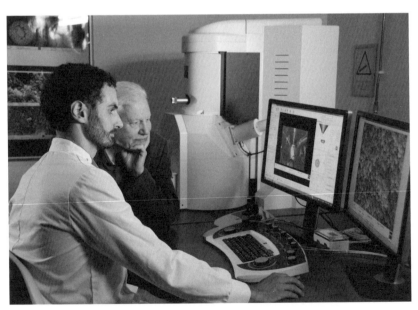

A typical day in the SEM lab in Bristol: David Attenborough
pops by in 2017 to see Fiann Smithwick at work.

preparators do not have to risk damaging delicate specimens, say a dinosaur embryo inside its egg, instead capturing a perfect 3D image.

CT scanning of fossils has only become commonplace in the twenty-first century, when scanners, developed first for medical use, became cheap enough that every university or museum could afford one. We commonly scan fossils up to the size of a magnum bottle of champagne; above that, and they have to go to industrial or veterinary scanners designed to scan an aircraft engine or a horse.

The image stacks provide the information for a 3D digital model. Generally with fossils irregularities in the rock complicate the picture, and students sometimes have to spend weeks editing the scan slices, digitally removing irregular rock grains, bits of fossil shell, and other debris. They can also colour code the different elements of the fossil, and then use the 3D model for further experiments to test, say, its engineering properties in feeding or locomotion.

Other advanced technological applications can also be applied to dinosaurs. For example, in our quest to identify the colour of dinosaur feathers, we used a scanning electron microscope, which enables scientists to see structures that are much smaller than can be viewed under a regular optical microscope. A light microscope allows scientists to see objects down to one-thousandth of a millimetre across, whereas a scanning electron microscope takes this to one millionth of a millimetre. We also use the scanning electron microscope to map the chemicals present in fossil bones or feathers, showing whether they were preserved as calcium phosphate or clay minerals, or were enriched in any other chemicals, such as iron or copper, that might give a clue to the mode of preservation. Palaeontologists now use the latest mass spectrometers, instruments that can identify inorganic and organic chemicals, even in tiny quantities, and which are becoming essential in the study of colour and the survival of any organic materials in dinosaur fossils.

How do we see the whole animal?

Once the bones have been collected and brought back to the lab, there are two further steps. First, the skeleton can be mounted for show in a museum; and second, its living form can be reconstructed by restoring soft tissues such as muscles, sense organs, and skin.

Once the bones have gone to a museum, the skeleton is built up in its correct configuration using a metal framework called an armature, which is strong enough to hold the bones, and shaped in such a way that the whole skeleton is arranged correctly and posed in a reasonable way. Some aficionados used to like to hide the armature inside the fossil bones, so great holes had to be drilled through the vertebrae so they could be threaded on to it, like half-tonne cotton reels. Now, every effort is made not to damage the bones, and indeed the armature may be visible.

How can the skeleton be put together correctly? We all know the kids' movies and cartoons in which the bones are strung together randomly, and perhaps the head is popped on the end of the tail. Well, in many cases, as in the excavation I did in Dinosaur Provincial Park, the skeleton is preserved more or less unperturbed, and with all bones in the right places. In any case, palaeontologists are like surgeons – they can immediately identify what each bone is – left femur, right humerus, dorsal vertebra, and so on. This is what they are trained to do. If anything is missing, the museum technicians can make casts of some ribs or vertebrae from their neighbours, or they can flip a right femur, say, to make a left femur. The symmetry and repeatability of the skeleton mean it's pretty clear when they get it right or wrong.

In travelling dinosaur shows, the skeletons are usually casts, often in artificial materials such as fibreglass, which makes them light and tough – easy to transport in pieces from venue to venue. The museum technician first paints the original bones with a rubber compound to create moulds, constructs a tough supporting cradle, and then separates the two or more pieces of the moulds, releasing them from the bones. These moulds can then be used to produce as many casts as are required, and the casts will show every fine detail of the original bone.

How can flesh be put back on the museum bones? Normally, this is done through a conversation between the palaeontologist and the artist. The skeleton carries many clues to the location and nature of soft tissues. For example, muscles generally attach to the bones at each end – the biceps muscle in the arm, which body-builders like to show off, attaches to the shoulder blade and to the main bone of the forearm, the ulna. This muscle, and indeed most of the other main muscles of the arms and legs, are pretty much the same among mammals, birds, and crocodiles, so they were probably comparable in dinosaurs. The sites where muscles attach to the bones often leave clear rough patches, and these can be used to reconstruct the angles and sizes of muscles in a dinosaur.

Muscles, skin, eyes, and tongues are put in place using any clues that exist on the bones, but otherwise by comparison with modern animals.

Later chapters will reveal how modern palaeobiological studies have told us a great deal about how dinosaurs reproduced, grew, fed, and moved, and all this new knowledge is used by the artist in restoring their image of the living dinosaur – whether it's a painting, a 3D model, or an animation. We can now even restore feathers and colours in some cases.

How can dinosaurs be used in education?

Museums have a key function in education. All the effort and expense of digging up dinosaurs and bringing them back to the museum can be justified partly in terms of the science. In addition, a key duty for museums, and indeed a key duty for university professors, is to take their science out to the public. Here is one example of how the crossovers between practical work, science, and education can work together.

Over the past twenty years, the University of Bristol has run a programme called the Bristol Dinosaur Project. Since 2000, the team has visited hundreds of schools and spoken to tens of thousands of children, as well as appearing at science fairs in Bristol and elsewhere. The Bristol dinosaur, *Thecodontosaurus* (see overleaf), was named in 1836. It may not seem very exciting because it is known only from isolated bones, and it's quite a small plant-eater, not much larger than an eight-year-old

Children look at the Bristol dinosaur
Thecodontosaurus at an educational festival.

child. Nevertheless, children love to hear about the dinosaur that stomped around their city some 208–201 million years ago, in the Late Triassic.

We use the dinosaur as a way to get children of all ages thinking about key science topics, such as geological time, continental drift, climate change, evolution, and biology. It helps enormously that we talk about the sciences, as in this book, in terms of testable ideas. Then the kids can follow through the calculations, say, of dinosaur running speed, and they see how it makes sense.

In the initial years, when we had a full-time Dinosaur Education Officer, we visited 200 schools and spoke to 10,000 children each year. Then, the Bristol Dinosaur Project received substantial funding from the UK Heritage Lottery Fund, and it was able to operate at a much more ambitious level; after the funding ran out, we had to scale back, but continue with enthusiasm. Our students love having the chance to try out their teaching skills, and to talk to young enthusiasts about what they love to do.

Genus: **_Thecodontosaurus_**

Species: _antiquus_

We speak to two age groups, seven- to nine-year-olds, and fourteen-to fifteen-year-olds, and the style is different with each. With younger children, it has always been very easy to engage their enthusiasm – they just love it when you pass round a dinosaur bone or tooth, and they are thrilled to be handling the real thing. With the teenagers, it is more important to let them see how much fun a career in science, or a related field, can be. We use a kind of forensic approach – here's a mystery that seems impossible to solve ('how fast could *T. rex* run?' or 'which month did the asteroid that wiped out the dinosaurs strike?'), and then step the students through the evidence and the basic theory they need. We are trying to engage them in science, showing them it can be fun, and why they need to study maths, biology, chemistry, and physics if they want to pursue it as a career.

The Bristol Dinosaur Project visits schools, but does many other things. Twice, it has been associated with outdoor exhibitions of

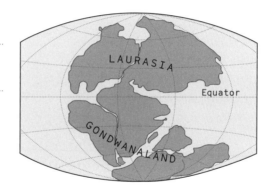

Named by:	Henry Riley and Samuel Stutchbury, 1836 (genus); John Morris, 1843 (species)
Age:	Late Triassic, 208–201 million years ago
Fossil location:	England
Classification:	Dinosauria: Saurischia: Sauropodomorpha
Length:	1.2 m (4 ft)
Weight:	40 kg (88 lbs)
Little-known fact:	This was the first dinosaur ever to be named from the Triassic.

animatronic dinosaurs at Bristol Zoo; these attracted tens of thousands of visitors. We work with Bristol City Museum and other museums, providing the enthusiastic personnel to talk to visitors; our students can answer questions in a personal way and use their own experiences to give down-to-earth answers.

The Bristol Dinosaur Project has also been a great vehicle for giving undergraduates their first research experiences. We can't offer them dinosaur bones for study, but there are associated projects, particularly on microvertebrates, such as the teeth of sharks or tiny bones of other fishes and reptiles. There are many cliffs and old quarries around Bristol that have yielded fossil-rich rocks, and we focus on concentrations of bones both in bone beds laid down on the seabed and on cave fills that include fossil bones. The students love the chance to do fieldwork, and they have to train their eyes to pick out the tiny bones and the bone beds in the rock sections.

The biggest challenge for the students is to organize their work into the correct, professional style for scientific publication. This is a steep learning curve, but so far twenty-five students have taken the projects to completion – and the publications help them advance their careers. Five or six successful palaeontologists can trace their careers back to such early experiences. This would seem to be a natural endpoint of the process we have traced in this chapter, all the way from digging up the bones to learning something new from them.

Digging up dinosaurs is one of the best things a palaeontologist can do. The field methods haven't changed much for over 150 years – nothing beats a good pair of eyes, and some strong shoulders! Three things have improved about the fieldwork. First, it's quicker and easier to travel around the globe, and so palaeontologists can now work in many parts of the world; this has the wonderful side-effect of encouraging more collaboration between young scientists from different countries than ever before. Second, we observe more, and especially about the context of the bones. This provides key sedimentological data for interpreting the ancient environments in which the bones have been preserved. Third, and associated with this, we are better at mapping the sites and better at recording the finds and making sure that nothing is lost, especially tiny fossils of fishes, frogs, or lizards that lived under the feet of the dinosaurs.

In the laboratory, many of the techniques of preparation – removing the bones from the rock – are also classic and have remained unchanged for a century. Today, we do have better power tools, better chemicals, and perhaps better respect for the fossils, so that we try to minimize damage. The study of fossil specimens has been revolutionized, though, by new equipment such as CT scanners and scanning electron microscopes. These have opened up possibilities we would not have dreamed possible a decade or so ago.

Digging and cleaning up dinosaur bones can be great fun, but of course it is only really worthwhile if we can then use those bones to learn something about how the dinosaurs lived. In the next chapter we explore whether dinosaurs were warm-blooded or not, how they breathed, and whether they were as stupid as they are supposed to have been.

Chapter 4

Breathing, Brains and Behaviour

Bringing dinosaurs back to life might seem a vain pursuit, and yet that is what palaeobiologists attempt to do. When I was an undergraduate, back in 1975, I was aware of a huge controversy about dinosaurs – were they warm-blooded or not? This engaged many smart scientists, and the public of course. The reporting often focused on the heated arguments – and the various names the scientists called each other, but then everyone likes a bit of rude behaviour. The scientific community had chosen a core question about dinosaurs, and one that has proved difficult to resolve.

The story goes back long before 1975. About 1840, when Sir Richard Owen began his studies of dinosaurs, and indeed when he was gearing up to name the group in 1842, as we saw in Chapter 2, he was also thinking about their palaeobiology. His core question was 'where did the dinosaurs fit into the story of life?' At the time, speculation about evolution was generally regarded as dangerous or improper, perhaps something for the French philosophers to contemplate...but the English elite were aware of what that sort of thinking led to – the French Revolution of 1789 – and didn't want any of that sort of nonsense on their side of the Channel!

Nonetheless, Owen was a skilled anatomist, and the evidence was there. Plants and animals showed evidence of shared similarities, but often with different functions, what Owen and we call homologies. A homology is an anatomical feature with a fundamental structure that is shared, but adapted, by different creatures – such as the arm of vertebrates. We know that the arm of the bird is modified as a wing, the arm of the whale is modified as a fin, the arm of a horse has a single finger and a single hoof, and the human arm has five fingers, but they are all homologues because the fundamental structure is the same: a single upper arm bone (the humerus), two forearm bones (ulna and radius) and fundamentally five fingers can be identified in the bird wing or whale's fin.

Owen must have struggled to avoid an evolutionary interpretation – that these arms are homologous because they shared a common ancestor. Had they been created, there would be no need for them to show the

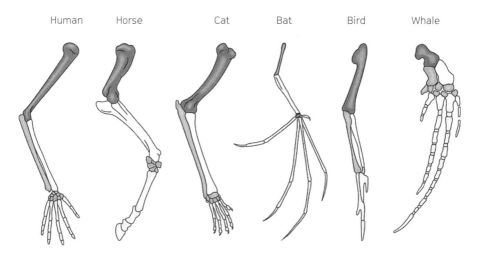

| Human | Horse | Cat | Bat | Bird | Whale |

Homology of limbs of six different vertebrates.

identical arrays of bones deep inside; but it all makes sense if they shared an ancestor. Later, Owen was framed as the opponent of evolution after the publication of Charles Darwin's *On the Origin of Species* in 1859.

For Owen, the dinosaurs were amazing and tricky. He looked for similarities with modern reptiles, and found a few. Still, he could see they were not overgrown crocodiles or lizards, as others had said before, and he had the insight and the courage to identify them as members of an entirely new group: hence the name Dinosauria. Further, and amazingly, he argued that they were mammal-like in many characters, including that they were probably warm-blooded. When he, crusty as he might seem to us now, was commissioned to advise on the London Great Exhibition of 1851, he was frivolous enough to imagine the first serious reconstructions ever of dinosaurs, the famous Crystal Palace models of 1853 – he showed *Iguanodon* and *Megalosaurus* as overgrown rhinoceros-like animals.

Owen had his reasons. He wanted to argue against evolution by showing that the ancient reptiles were more advanced than modern reptiles, that reptiles had in some way degenerated over time. Be that as it may, we can be grateful to the old warthog for giving us the name 'dinosaur', and for daring to step out of line and do the daringly populist thing of bringing whole-body reconstructions to the public – he triggered the first phase of 'dinomania', a term invented later to describe the all-consuming public appetite for dinosaurs.

Most importantly, because of his great seriousness and prestige, Owen could sanction such speculation and be accepted. From 1842, palaeontologists and others have sought ways to get to the heart of

Sir Richard Owen posing with a friend.

Megalosaurus (foreground) and *Iguanodon*
as visualized by Richard Owen in 1853,
in Crystal Palace Park, London.

dinosaurian physiology – how much they ate, how they powered their bodies, and whether they were warm-blooded or not. The quest has involved experts from many fields, new techniques such as the study of fine-scale bone structure, and remarkable new fossils such as the feathered dinosaurs of China.

Were the dinosaurs warm-blooded?

So, were the dinosaurs warm-blooded or not? The answer is, as ever, yes and no. When I joined the debate, with a cheeky article I wrote as an undergraduate, and which was published in 1979 in *Evolution*, the leading American journal in the field, palaeontologists saw things as polarized. A dinosaur was either warm-blooded like a bird or mammal, or cold-blooded like a reptile. The term 'warm-blooded' is a slight misnomer, because the trick birds and mammals have perfected is to keep their internal temperature *constant*, not necessarily warm; it just so happens that most biologists operate in temperate climates, so when they grab a dog or a baby or a chicken, it feels hot.

This constant temperature comes at a cost, however – typically a human or dog that weighs the same as a crocodile has to eat *ten times as much food*, because nine-tenths of what we eat is used simply to regulate our core temperature. This is why the crocodile lazes about so much, grinning sardonically at us – he has a great secret. If warm-bloodedness is so costly, why do it? The reason is that birds and mammals can be active all day and all night (lizards and crocodiles are torpid in cold conditions, such as night time), and they can occupy cold parts of the Earth.

In my 1979 article, I picked up two points that were relevant. First, warm-bloodedness is not always *better* than cold-bloodedness, and second, that living animals show a gradation between the two states. Some new physiological work in the 1970s had shown that insects and reptiles could generate internal heat – think of the bumblebee flying on a frosty day, shivering like mad to warm up the flight muscles before taking off. Further, small birds and mammals often switch off at night because they can't eat enough to keep warm all the time; others hibernate, which is the same thing.

The debate was kicked off by maverick palaeontologist Bob Bakker, who, as a grad student at Yale in the 1960s, had first shown that many dinosaurs were fleet and fast, and then followed this through to its logical conclusion. He was a smart writer and skilled artist, and could conjure images of *T. rex* galloping through the undergrowth and the great sauropods rearing up on their hind legs to snatch leaves from high in the trees. Perhaps most would reject those images as over-fanciful, but Bakker stimulated the modern era of dinosaurian palaeobiology. People had to accept that dinosaurs did not have the physiology of modern crocodiles. First, many might have had feathers, especially those on the evolutionary line to birds, and what about the giant sauropods? How could a 50-tonne sauropod have functioned if it had the physiology of a modern lizard or crocodile? We need to explore these two ideas, and then see how the study of the internal structure of dinosaur bones helped to solve the conundrum.

Are birds living dinosaurs?

In 1984, Bob Bakker and Peter Galton, an English palaeontologist who became established in the United States, published a provocative paper in *Nature* entitled 'Dinosaur monophyly and a new class of vertebrates', in which they stated not only that dinosaurs were a single clade (see Chapter 2), but also that modern birds were dinosaurs: 'Recently Ostrom has argued forcefully that birds are direct descendants of dinosaurs and inherited high exercise metabolism from dinosaurs.' They were undoubtedly right, but this paper was designed to enrage the old guard.

John Ostrom was the real revolutionary, and he was Bob Bakker's doctoral supervisor at Yale. As with any professor from Yale at the time, Ostrom was reserved, polite, and always impeccably dressed, being famed for his brightly coloured plaid jackets. Ostrom had spent much of the 1960s excavating and describing an amazing new dinosaur, **Deinonychus** (see overleaf) from the Early Cretaceous of Wyoming. Ostrom could not escape the observations that, first, this dinosaur was built for speed and manoeuvrability as a hunter, with its huge slashing claw on the second toe of the foot; and, second, that the skeleton of *Deinonychus* was pretty well indistinguishable from that of **Archaeopteryx** (see overleaf), the first bird (see pl. iv).

When Ostrom published his monograph on *Deinonychus* in 1969, it was an instant hit – a very careful piece of anatomical description, with beautiful illustrations of an astonishing dinosaur. The frontispiece was an inspired pencil drawing of *Deinonychus* at speed, reflecting precisely John Ostrom's vision of the dinosaur, and a revolutionary depiction of dinosaurs as fast and active. The artist? Bob Bakker.

John Ostrom, affable as ever, but not wearing his signature plaid jacket.

The amazing image of *Deinonychus* drawn by Bob Bakker in his student days.

What is the evidence that birds are dinosaurs? Evidence was first noted back in 1870, in fact, by Thomas Henry Huxley, when he wrote about the newly discovered fossil bird *Archaeopteryx*. The skeleton had been unearthed in a limestone quarry at Solnhofen in southern Germany in 1861, igniting a bidding war among the museums of Europe. The fossil ended up in the British Museum in London, purchased thanks to the drive of Richard Owen, director of the natural history portion

Genus:	*Deinonychus*
Species:	*antirrhopus*

Genus:	*Archaeopteryx*
Species:	*lithographica*

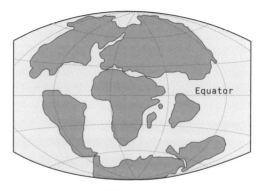

Named by:	**John Ostrom, 1969**
Age:	**Early Cretaceous, 115–108 million years ago**
Fossil location:	**United States**
Classification:	**Dinosauria: Saurischia: Theropoda: Maniraptora: Dromaeosauridae**
Length:	**3.4 m (11 ft)**
Weight:	**97 kg (214 lbs)**
Little-known fact:	***Deinonychus* preyed on the much larger *Tenontosaurus* either by slashing or biting into its flesh until it bled to death.**

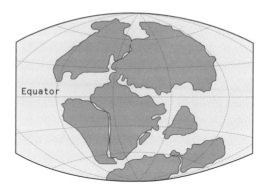

Named by:	**Hermann von Meyer, 1861**
Age:	**Late Jurassic, 152–148 million years ago**
Fossil location:	**Germany**
Classification:	**Dinosauria: Saurischia: Theropoda: Maniraptora: Avialae (birds)**
Length:	**0.5 m (1¾ ft)**
Weight:	**0.9 kg (2 lbs)**
Little-known fact:	**The first *Archaeopteryx* fossil to be found was an isolated feather, in 1860. The first complete skeleton was discovered a year later.**

of the museum, for the enormous sum of £700 (equivalent to £80,000 today). Owen wanted to have the specimen so he could publish the first description, which he did. But it was an embarrassment to him in many ways. He noted the close similarity of all its bones to those of dinosaurs, and indeed modern birds. He also noted the clear impressions of feathers on the wings and over the body.

Owen was loath to call *Archaeopteryx* a 'missing link', as he had opposed Darwin's dangerous new ideas of evolution, published two years earlier in 1859. Thomas Henry Huxley, by contrast, had no such qualms. As skilled an anatomist as Owen, he got sight of the specimen and used Owen's description to write his own paper about dinosaurs and birds. He pointed out all the similarities, and that *Archaeopteryx* was key evidence for evolution – the perfect intermediate between dinosaurs and birds, with its primitive long bony tail and teeth and its advanced feathers and wings.

For nearly a century, everything had seemed settled – additional specimens of *Archaeopteryx* and of small theropod dinosaurs continued to confirm Huxley's insights – but then the research field veered wildly off track. For all sorts of reasons, palaeontologists stopped seeing birds as dinosaurs – maybe they couldn't believe such amazing flying machines as birds could have evolved in as little as 20–30 million years, or they were afraid to admit that they had some great evidence of evolution in their hands. Whatever the reasons, it took a century for palaeontologists to emerge from their state of denial and to accept that Huxley was right in 1870, just as Ostrom was right in 1970: birds really are dinosaurs.

Ostrom noted everything Huxley had seen, and especially the fact that *Deinonychus* had bucked the trend of evolution of the other theropods – it was relatively small and it had long arms. Other theropods, such as *T. rex*, became huge and their arms dwindled. What Ostrom didn't know, but guessed, was that *Deinonychus* had had feathers, and indeed its arms were long just so that they could carry the specialist flight feathers seen along its arm and that of *Archaeopteryx* and other birds. Confirmation had to await the discovery of the remarkable birds and dinosaurs from China in the mid-1990s, as we shall see.

Ostrom, though, saw that theropods shared hollow bones with birds, as well as the fused clavicle (commonly called the wishbone) in the chest region, the semilunate carpal in the wrist (allowing *Deinonychus* and birds to fold the hand back, as birds do when they tuck their wings back along the side of the body), expanded eyes capable of 3D vision, an appropriately expanded brain (needed for leaping or flying from tree to tree), and many more characters.

Thomas Henry Huxley – did he perhaps
know how smart he was?

Nonetheless, since these early papers by Ostrom, Bakker, and Galton, there has been a remarkably vocal crew of nay-sayers who continue to express a counter view until well into the twenty-first century, and will doubtless carry on doing so. They have survived on the 'balanced' airtime given by scientific documentaries – 'here's one view; and here's the other'. Never mind that the bird-dinosaur view is supported by hundreds of independent bits of evidence and the 'birds are not dinosaurs' view lacks an alternative theory and lacks evidence. This might be the one negative aspect of the great public interest in advances in dinosaur science: the fact that proponents of rejected views can promote their ideas directly to the public even if the scientific journals, with their systems of scrupulous peer review, no longer accept their papers.

Bone histology and being huge

Ostrom's evidence that *Deinonychus* was a dinosaur close to the origin of birds in the evolutionary tree gave credence to Bakker's (and Richard Owen's) view that dinosaurs were warm-blooded. However, those early debates around 1970 were quite unsophisticated. Many of the lines of evidence brought forward at the time were suggestive, but not decisive, and so the debate meandered inconclusively. One, though, has proved to be fruitful.

This is bone histology, the study of the internal microscopic structure of bones. Since the 1800s, biologists had used the light microscope to study cells and microscopic life. Sections of bones showed their complex internal structure, with dense bone on the outside, and often more open bone tissue near the centre. In life there are no spaces, and bone is full of fat, blood vessels, and nerves. Bone histologists noted that modern cold-blooded animals in particular, such as fishes and reptiles, have distinctly layered bone – this tracks their fast growth in summer and slow growth in winter, and the growth layers build up more or less like growth rings in a tree. As we shall see later (Chapter 6), palaeontologists can use the growth lines to age dinosaur skeletons and to build up growth curves for individual species, showing how the rate of growth can vary over the years, from hatchling to adult.

Warm-blooded animals such as birds or mammals, on the other hand, tend to have bone without obvious layering, because they grow evenly all the time (a consequence of having internal temperature control) and the bone often shows evidence of remodelling. Bone remodelling is represented by tubular structures that cut through the background structure, and is a result of the high metabolic rates of birds and mammals, which mean that they lay down calcium and phosphorus in their bones, but also need to remobilize those elements at times for other purposes such as producing eggs or surviving a tough winter.

It turns out that dinosaur bone structure is more bird- or mammal-like than reptile-like. In the microscopic section illustrated opposite, there is a background of regular so-called fibrolamellar bone that has some layering, but this does not represent annual increments. The black spots are cavities in which the cells that build up bone and the cells that break down bone would have resided. Scattered over the field of view are secondary remodelled canals, some highlighted by orange iron staining – these cut across the regular structure. So, dinosaur bones show evidence of extensive internal remodelling, and Bakker rightly interpreted this to mean that dinosaurs were warm-blooded. As many pointed out at the time, though, there are ways and ways of being warm-blooded, and size is one of them. Some large crocodiles and snakes today show gigantothermy, a great word that says it all: they are huge and their size helps regulate their internal temperature. It's simple physics that if a cylinder is heated up, it cools fast if it is small, and takes much longer to cool down if it is large.

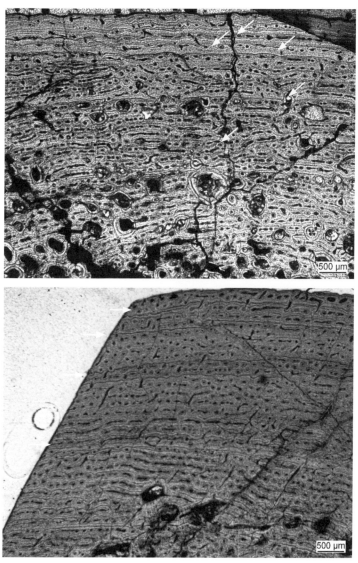

Bone histology of the dwarf sauropod dinosaur
Europasaurus, showing growth lines (white arrows).

So, in a series of wonderful experiments in the 1940s, Ned Colbert and colleagues showed that the core temperature of small alligators is more or less in tune with air temperature, but as the alligators got larger and larger, the hot-cold-hot cycles of day-night-day temperature did not directly drive the animal's core temperature; rather, it was *damped*, meaning that it adjusted up and down more slowly. The experimenters predicted that at a certain size the alligator would keep a constant core temperature, even though it was living in air temperatures that rose and fell by 20 or 30 degrees each day and night. I have always wondered how they determined the core temperature of those over-heated alligators – presumably using a thermometer on a very long broomstick!

Birds and crocodiles have a second, related, characteristic, almost certainly also shared with dinosaurs: they breathe air through their lungs in a one-way manner. Humans and other mammals breathe in and out using a *tidal* system, meaning that there is always some foul air in our lungs even when we puff out as hard as we can. Birds and crocodiles, on the other hand, breathe oxygenated air into the lungs, where oxygen passes into the bloodstream, and the air passes also into extensive air sacs around the backbone and guts. Then, when birds breathe out, everything is cleared from the air sacs and lungs. Dinosaurs, including sauropods, did the same, and this gave them a more efficient way to keep their metabolism high without eating huge amounts.

These, then, are two of the characteristics that perhaps allowed dinosaurs to be truly gigantic, and we will explore these further in Chapter 6. One-way respiration increased their ability to acquire oxygen and so to power a high metabolic rate with less energy than we have to use; and gigantothermy meant they could be warm-blooded just by being huge.

Mesozoic birds from China

We left the story of bird evolution with the recalcitrant dinosaur-bird deniers. In fact, and as if Huxley and Ostrom required any vindication, the discovery of abundant *feathered* dinosaurs from China since 1990 has been an amazing confirmation that birds are dinosaurs. I remember seeing the first images of feathered dinosaurs to reach the West at the annual meeting of the Society of Vertebrate Paleontology in New York in 1994. Two Chinese professors were there, wearing smart suits, and they created quite a stir – this was early in the days of the political opening

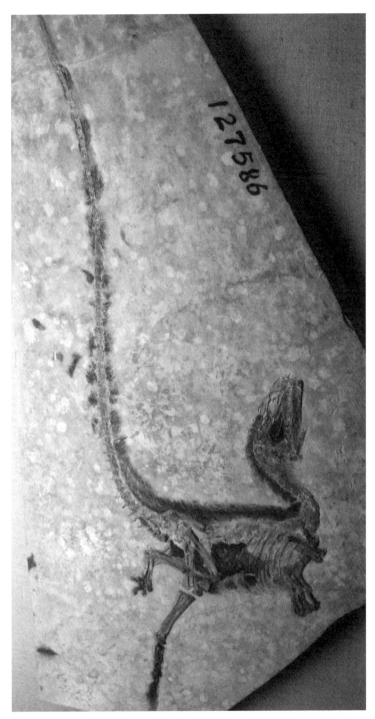

Skeleton of the first of the Chinese feathered
dinosaurs to be announced: *Sinosauropteryx*.

up of China, and we remembered how China had been a closed country. The professors had astounding photographs of feathered dinosaur fossils – there was the skeleton laid out, complete with traces of internal organs such as the liver inside the rib cage, and – there could be no doubt – a fuzz of bristly feathers round the edges.

These rare visitors from China – *rara aves,* one might say – were buttonholed by all the great and the good. Shortly after, John Ostrom,

Skeleton of the four-winged gliding dinosaur *Microraptor.*

Phil Currie and colleagues visited China for the first time, and they were convinced of the fossil's authenticity. The beast was named in 1996 by Drs Ji and Ji, and the images made available for all to see two years later in a fuller description by Pei-ji Chen and colleagues, in the world's leading scientific journal, *Nature*. This was **Sinosauropteryx** *prima* (see overleaf); little did I realize at the time that I would one day have a chance to study it.

The critics declared it was a fake, artfully put together from bits and pieces of several skeletons, and with feathers glued on. Those who had seen it knew it was real. Professor Chen and colleagues were cautious in their *Nature* paper, however, and called the feathers 'protofeathers', saying that 'much more work needs to be done to prove that the integumentary structures of *Sinosauropteryx* have any structural relationship to feathers'. This caution was understandable, but soon the specimens piled up and the feathers were unequivocal. Whereas in *Sinosauropteryx*, the feathers were just bristles, in **Caudipteryx** (see overleaf), named in 1998, there were branching feathers, like the down feathers of a modern bird. Then *Microraptor*, named in 2000, showed all the flight feathers you could wish for – primaries and secondaries arrayed along the wing. And, not only that, they were arrayed along the hind wings. This was a four-winged beast, similar to a postulated 'tetrapteryx' or four-winged flyer that some experts on the origins of flight had earlier speculated must have existed.

Here was a dinosaur, with a wingspan of less than 1 metre (3 feet), that could fly, but not exactly like a bird – well, not at all like a bird. *Microraptor* was in fact a close relative of Ostrom's *Deinonychus*, a member of the Dromaeosauridae, which were close to the origin of birds. *Microraptor* has been reconstructed and modelled by aerodynamics experts. It might have flown like a kite, with both sets of wings on the same plane, or like a World War I biplane, with the front wings above the hind wings. Either way, it almost certainly used its wings in leaping and gliding from tree to tree, not in flapping flight. The area of the wings is not sufficient to support the body mass during prolonged flights.

Therefore, in evolutionary terms, the new Chinese fossils have shown that the origin of birds, far from being the sudden event people had speculated, was a long and complex process. The early palaeontologists may have rejected the bird-dinosaur model because they thought a flying bird could not have evolved rapidly from a great lumbering theropod dinosaur like *Allosaurus* or *Tyrannosaurus*. And they were right. The creationists love to pick on *Archaeopteryx* as the great 'missing link' fossil – if you can ridicule *Archaeopteryx* (little fluffy bird hatches out of crocodile egg), then you can claim to destroy evolution.

Genus:	**_Sinosauropteryx_**
Species:	_prima_

Named by:	**Qiang Ji and Shu'an Ji, 1996**
Age:	**Early Cretaceous, 125 million years ago**
Fossil location:	**China**
Classification:	**Dinosauria: Saurischia: Theropoda: Compsognathidae**
Length:	**1 m (3 ft 3 in.)**
Weight:	**1 kg (2 lbs 3 oz)**
Little-known fact:	**This was the first dinosaur to have its feather colour determined, in early 2010.**

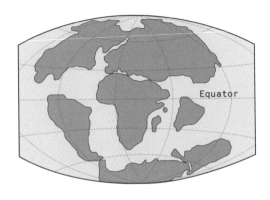

Chapter 4

Genus:	*Caudipteryx*
Species:	*zhoui*

Named by:	Qiang Ji and colleagues, 1998
Age:	Early Cretaceous, 125 million years ago
Fossil location:	China
Classification:	Dinosauria: Saurischia: Theropoda: Oviraptorosauria
Length:	1 m (3¼ ft)
Weight:	1 kg (2 lbs 3 oz)
Little-known fact:	At one time, it was suggested that *Caudipteryx* was a flightless bird, but it is clearly a theropod and not a bird.

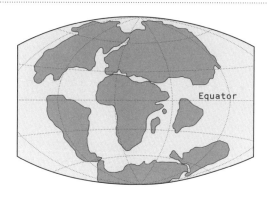

We now know, thanks to the Chinese fossils from the Jurassic and Cretaceous, that there were dozens of feathered, flying dinosaurs, all experimenting with different styles of gliding and parachuting. Then one lineage, of which *Archaeopteryx* is an early representative, cracked true powered flight, in which the wings beat up and down. That gave them the breakthrough to success, with hundreds of bird species flourishing in the Cretaceous, and nearly 11,000 species today.

Can we tell the colour of dinosaurs?

I discussed the topic of dinosaur colour in the Introduction, and this has been one of the most exciting and unexpected recent discoveries in dinosaurian palaeobiology. Unexpected because we used to lament, 'We'll never know what colour they were'. It might be plausible to reconstruct feeding and locomotion from the bones, but surely colour would require a time machine?

As I noted in the Introduction, however, the secret is that much of the colour in bird feathers and mammal hairs comes from variants of the pigment melanin. One form of melanin, called eumelanin, gives all the black, brown, and grey colours, and another, phaeomelanin, gives ginger colours. This is all that mammals have, whereas birds have two other pigments in their feathers, porphyrins to give purple and green colours, and carotenoids to give red and pink colours. The key is that melanin is a very tough chemical that survives a great deal of heat or compression, and so it survives in fossils. Further, the two key types of melanin are contained within differently shaped capsules, called melanosomes – sausage-shaped ones for eumelanin, spherical for phaeomelanin – and this is constant in birds and mammals. Therefore, applying the extant phylogenetic bracket idea (in evolutionary terms, mammals and birds 'bracket' the dinosaurs), it is likely that the same shape–colour relationship applies to all included groups, including dinosaurs. Melanin is produced in the skin and passes into melanosomes in the developing hair or feather through the follicle.

I first had the chance to go to China in 2007, with colleagues Paddy Orr and Stuart Kearns. We spent two weeks in the field, exploring all the localities in northeast China's Jehol Beds, a major set of formations of Early Cretaceous age, that had yielded specimens of feathered birds and dinosaurs, and a further two weeks in the laboratories of the Institute of Vertebrate Paleontology and Paleoanthropology (IVPP) in Beijing. There, we peered down our microscopes at samples of feather and skin and spotted some promising examples.

When we read the important paper in 2008 by Jakob Vinther, then a doctoral student at Yale, who had identified melanosomes preserved in fossil bird feathers from Brazil and Denmark, we immediately thought, 'let's see if we can find these in dinosaur feathers'. We spoke to Fucheng Zhang at IVPP, who had visited Bristol in 2005 to work on fossil bird specimens, and arranged to borrow some samples, including small slivers of fossilized feathers from different parts of the body of *Sinosauropteryx*, and he visited Bristol for a second time in 2008. That was when we found the melanosomes.

We wrote the paper and sent it to *Nature* in early 2009. As is the way of things, it took forever to convince all the reviewers. Indeed, the paper was reviewed twelve times – three cycles of four reviewers each time – and one was absolutely not to be convinced. 'They're not melanosomes, these aren't feathers, and they aren't dinosaurs...' I was able to talk to Vinther and colleagues while on sabbatical at Yale in early 2009, and eventually our paper came out in February 2010. We showed that *Sinosauropteryx* had phaeomelanosomes, the capsules that contain the ginger version of melanin, and lots of them. It was ginger! The tail was striped, with equal bands of white and ginger along its length. So, we published our reconstruction (see pl. v) with a confident statement: 'The reconstruction shows the correct colours of a dinosaur for the first time.' This is important: we were not giving an opinion, however informed, but stating an objective fact, and our statement could be refuted if someone showed our observations of melanosomes to be false.

At the same time, the Yale team, led by Jakob Vinther, published their reconstruction of an even more gaudy dinosaur, *Anchiornis* from the Jurassic of China, which sported black and white stripes on its wings and tail, and a lovely ginger crest on top of its head, as well as specklings of black and ginger feathers on its cheeks. What does this all mean, though? Determining the colour of a dinosaur may be a bit of smart lateral thinking, perhaps a clever parlour trick, but can it tell us anything useful?

Did dinosaurs indulge in sexual selection?

Identifying the colour of dinosaur feathers revolutionized our appreciation of the complexity of dinosaurian behaviour. Birds today have feathers for three main reasons – insulation, signalling, and flight.

It was obvious that insulation came before flight; the downy feathers over a bird's breast are to keep in the warmth and help its thermoregulation, and these feathers are much simpler than the flight feathers. So, it was assumed that if dinosaurs had had feathers, as Bakker suggested, they would likely have been for insulation. However, in our 2010 papers, both our team and the Vinther team declared that feathers were clearly for signalling from an early point in their evolution. We could not, however, go so far as to say that had been their initial purpose – but it might have been.

The striped tail of *Sinosauropteryx* and the barred wings and coloured crest of *Anchiornis* could have had no other function than signalling. No such patterns would be needed for insulation or flight. Further, it does not look as if the colours are there for camouflage – the barred tail might suggest it, but animals today that use bars for camouflage, such as tigers and zebras, have bars all along the body, not just on the tail.

So, signalling means sexual signalling. We can now imagine male dinosaurs, particularly the small theropods, as hopping about and showing off their wares to the females, just as so many birds do today. One reason that birds are so diverse, with nearly 11,000 species known, is that sexual selection helps to maintain and drive splitting of species, each with its own colourful feather patterns. Strip off the feathers, and most perching birds have nearly identical skeletons, for example, but when the males are equipped with their plumage, they are gloriously different, and the species do not interbreed because their pre-mating dances and displays are intensely interesting only to females of that species.

The realization that many dinosaurs might have been sexually selected has posed a conundrum: not many of them show evidence of sexual dimorphism (differences in form between male and female). Today, many reptiles, birds, and mammals show sexual dimorphism – think of the sleek lioness and the larger, maned male lion, or many examples among primates in which the male is much larger and equipped with bigger teeth. Maybe, though, birds give part of the answer – although the male and female peacock are indisputably different, it's all in the feathers. Their skeletons are very similar, perhaps differing in some small details of size. The same might have been true of displaying theropods too.

This has been part of a recent heated debate between those who see many horns and crests in dinosaurs as evidence for sexual dimorphism or sexual signalling, versus those who would interpret all such structures as having different functions, for example in feeding, defence, or species recognition. Kevin Padian and Jack Horner made a strong case for the

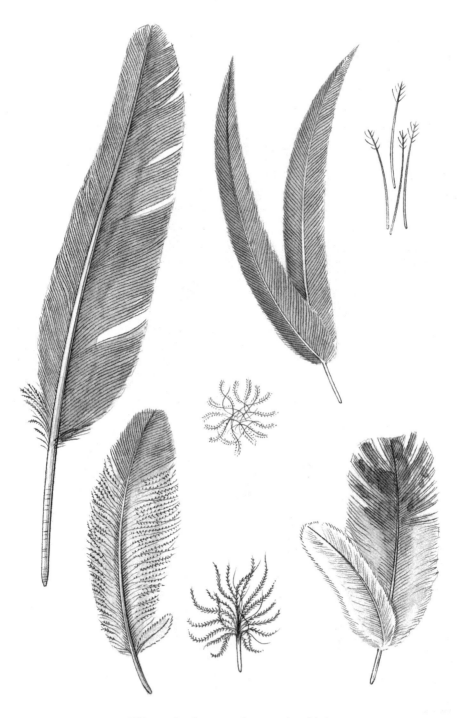

Different feather types from modern birds.

'species recognition hypothesis' in a paper in 2011 – they argued that all 'bizarre structures' in dinosaurs were to allow individuals to spot other members of their own species, perhaps across a crowded landscape containing many other similar-looking dinosaurs, for reasons of mutual protection. In such a model, sexual selection would be of minor importance among dinosaurs.

In a direct riposte, Rob Knell and Scott Sampson argued that species recognition might be a secondary function of the horns, crests, and feather arrays of many dinosaurs, but that the only valid argument to explain the costliness of evolving and maintaining such structures is sexual selection. Further, they noted, the shapes and extents of the bizarre structures are so variable among members of a single species that they might not have been very useful as unequivocal labels of species identity, but rather that they were under selection for other functions including mate competition, weaponry for fighting other males, or ornaments for showing off to females.

The debate rumbles on, but all of the evidence for quite complex social behaviour in dinosaurs suggests that they were perhaps not so stupid as they have been portrayed.

Were dinosaurs brainy or not?

Are birds (and dinosaurs) intelligent or not? Sexual display, and all the complex behaviours often associated with it, might imply high intelligence. And yet we say 'bird-brained' to mean stupid, and although birds have high-domed skulls and bulging little brains, much of the brain tissue is given over to powering their exquisite senses, especially sight.

It's also a given for many that dinosaurs were stupid. They are portrayed in museum displays and kid's books as mindless automata that barged around knocking over Jurassic trees, and surviving only because all the other dinosaurs were equally stupid. Famously, we learn that the plate-backed *Stegosaurus* had a brain the size of a kitten, and that it even had a larger brain in its hindquarters to control its tail and back legs.

Humans, and mammals in general, owe their claim of intelligence to the cerebral hemispheres, the forebrain, the two great spheres of wrinkled brain tissue we see when a brain is presented in the surgeon's hands. These wrap around the so-called 'primitive' regions of the brain, the midbrain and hindbrain. In fishes and reptiles, the brain is more linear, with the hind-, mid-, and forebrain regions in a row. In simple terms, most of the

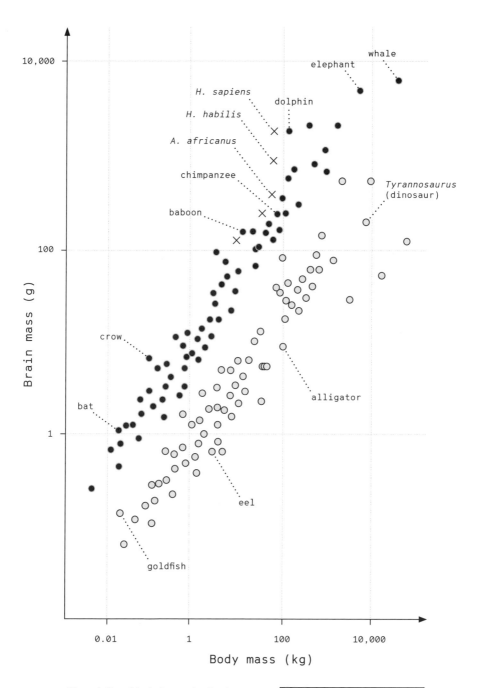

The relationship between brain size and body size for mammals, birds, and reptiles, including dinosaurs.

Legend:
- X Primates
- ● Mammals and birds
- ○ Fish and reptiles

reptile brain operates the sense organs – nose, eyes, and ears, as well as reflexes, and repetitive behaviours such as fight or flight and seeking food.

Dinosaur brains are not preserved, but their impressions are there deep inside the fossil skulls. We tend to think of the head as mainly full of brains, as it is in humans and other mammals. In reptiles, including dinosaurs, the brain is actually pretty tiny. It sits inside the braincase, a bony structure located deep within the skull, and the relative sizes are like a match box inserted inside a shoe box. Most of the dinosaur's head is full of jaw muscles at the back, and eyes and nasal cavity inside the snout.

Dinosaur neuroscientists used to look for natural rock casts from inside the braincase. Then they found that they could gain some understanding of the brain shape and size by filling up the braincase region of a cleaned-up fossil skull with casting medium. Now they use CT scanning, and the resulting brain models (see pl. xi) can be spectacular, showing the so-called 'optic lobe', responsible for processing images from the eyes, on two stalks sticking well forward, and then the mid- and hindbrain portions within the braincase. At the sides, the cranial nerves sprout out, those key nerves that make the organs of the face function. The semi-circular canals of the middle ear are even visible (see pl. xii) – these are the key organs of balance.

This is all very well, but were dinosaurs intelligent or not? Intelligence has always been a topic of great importance for humans because we define ourselves based on our intelligence – we called our species *Homo sapiens*, meaning 'wise person'. Indeed, our brains are large, but a whale's brain is larger. Is it more intelligent? Not necessarily, because of course brains are scaled in proportion to body size. So, brain biologist Harry Jerison proposed a measure of the ratio of brain size to body size, which he called the encephalization quotient (EQ) in 1973, and he argued it was a pretty useful way to measure animal intelligence. As expected, mammals have relatively large brains, whereas reptiles have relatively small brains. Birds fall somewhere in between, closer to mammals than to reptiles, and dinosaurs fall between modern reptiles and birds in their EQ.

So, dinosaurs on the whole weren't very brainy at all, even though some of them might have shown complex behaviour around mate choice. We can probably say that some dinosaurs at least, perhaps the small theropods, were as brainy as birds, and more intelligent than lizards or crocodiles. Here we are pushing the limits in trying to retrieve information about unfossilizable soft tissues, but could we find more? What about unusual conditions of preservation?

Can dinosaurs be preserved in amber?

Who could dream of a better headline than 'dinosaur preserved in amber'? And yet that's what happened in 2016, when we announced one of the most spectacularly perfect dinosaur fossils ever discovered. I had the good fortune to be invited by Lida Xing, palaeontologist at the China University of Geosciences in Beijing, to be part of a team to describe an amazing fossil he had acquired in 2016 – a tiny dinosaur tail preserved in amber (see pl. vi). Under the microscope, you could clearly see the bones of the tail skeleton, the fluff of feathers around the tail, and even the withered remains of tail muscles and skin inside the amber.

This was the most reported fossil discovery of 2016, with thousands of headlines around the world. In fact, it was rated as the eighth most reported scientific discovery that year, based on the number of reports, Tweets, and Facebook mentions. Look at the photographs – this is a truly astonishing fossil!

The specimen came from Myanmar (Burma), from famous mid-Cretaceous amber deposits that had been known for their fossils since the 1890s. In a review published in 2002, palaeoentomologist David Grimaldi reported beautiful specimens of angiosperm flowers and other plant remains, as well as examples of thirty families of insects and spiders, and some isolated feathers. By 2010, the number of insect families represented had risen to nearly 100. The amber is dated at 98.8 million years ago, placing it squarely in the early phase of the Cretaceous Terrestrial Revolution, the key event mentioned in Chapter 2, when flowering plants and all their buzzing accompaniment of insects exploded onto the Mesozoic scene.

Amber is a yellowish or orange-brown substance, partly transparent, and quite light in weight. It has been collected for centuries and used in jewelry and decoration. Many amber pieces contain insects and leaves, and these are often sold as unusual and attractive pendants and brooches. Amber is the fossilized resin of ancient trees, especially conifers such as pines and cypresses, which leak a sticky resin-like sap from their bark. Close study of the trapped organisms inside the amber shows all the fine details – for example, a tiny insect trapped in amber might show microscopic hairs over its back, and every lens in its compound eye. Some of the specimens even show colour patterns and perhaps the original colours. In the amber collections around the world, as well as insects and bits of plants, people have also found very rare examples of mushrooms, feathers, mammal hair, and even whole lizards and frogs.

Amber is known from many localities, including the Baltic and Dominica, and it is mainly mid-Cretaceous to Cenozoic in age, so only known from the past 125 million years. New papers about fossils from the Burmese amber are coming out at the rate of 100 or more per year, so eventually this one assemblage will probably number hundreds of species.

....

Understanding that birds are dinosaurs opened a new field of research, and the huge number of astonishing new fossils of feathered dinosaurs from China were perfect vindication of the great endeavours and insights of leading Victorian scientists such as Huxley and Darwin. New fossil finds fuel the field of palaeobiology, and the fossils from the Burmese amber in particular are providing access to specimens and soft tissues that would previously have been believed long destroyed.

Yet, as we have seen, it is not just new finds, but also new technologies, that increase our knowledge. In delving deep into the structure of dinosaur bones and feathers, using CT scanning and high-powered microscopy, we have learned more in the past decade about thermoregulation, colour, and behaviour than in the previous century.

Often, as we have found, the knowledge about modern organisms and tissues just isn't there, and so the dinosaur studies stimulate a new investigation of their living relatives. For example, when Vinther and we began our studies of feather colour, there was no compendium of the distributions of melanin and melanosomes among the differently coloured feathers of modern birds. This stimulated ornithologists to gather together stray natural history observations and construct a detailed data frame that links colour to structure and chemistry, and so enables the palaeontologists to interpret their fossils reliably.

The future will see more exceptional fossils, and closer attention to detail in exploring their microscopic structures. I still can't imagine a better fossil than our dinosaur tail in amber...unless it was a whole baby dinosaur?

Chapter 5

Jurassic Park? (Or Not...)

Traditionally, dinosaurs have been famed only for their extinction, but that's quite negative – certainly from the dinosaur's own point of view. Surely their lives were amazing, and we explore their astonishing palaeobiologies as living, breathing, eating, running, growing, and mating creatures in this book. How much fun it would be, then, to have a living dinosaur!

The idea has been posited many times in science fiction. In his 1912 novel *The Lost World*, Sir Arthur Conan Doyle describes how the zoologist Professor George Edward Challenger and his team explore remote regions of South America. They have heard rumours of a high plateau in the mountains, far from any civilization, that remains frozen in time from the days when dinosaurs roamed the Earth. After many adventures, the explorers reach the plateau, and they find a strange ancient world,

Sir Arthur Conan Doyle's *The Lost World* (1912)
was first made into a film in 1925.

populated with fiendish ape-men and terrifying prehistoric creatures. Challenger's team are chased by dinosaurs, and great leathery-winged pterodactyls dive-bomb them from the air. Eventually they get back to safety, and they bring a baby pterodactyl back to London to show to an amazed and disbelieving world.

Towards the end of the First World War, Edgar Rice Burroughs imagined a world of living dinosaurs and mammoths on the island Caprona, mysteriously located somewhere in the South Pacific, in his novel *The Land that Time Forgot* (1918), another classic in the genre. His story involves German and British troops, U-boats, and a world at war.

There were many more such adventure tales through the twentieth century, but the most convincing and scientific account was by Michael Crichton, in his novel *Jurassic Park*, published in 1990, and made into a film by Steven Spielberg in 1993. The story is well known, and founded in Crichton's knowledge of major advances in genomics around that time. He suggested that minute fragments of dinosaur DNA recovered from the gut of a blood-sucking mosquito preserved in 100-million-year-old amber could be amplified, inserted into the ovum of a modern amphibian as the host, genetically engineered, and then a baby dinosaur would hatch from the egg.

The focus on DNA (short for 'deoxyribonucleic acid') was justified, because DNA famously carries the genetic code – it is the stuff of the chromosomes inside the nucleus of every cell in your body. In a human, there are some 3 billion base pairs (bits of genetic code), distributed through 46 chromosomes (2 × 23 unique chromosomes), and these base pairs are arranged into 30,000 genes which, together, provide all the instructions to make a human being, and to maintain the cells by repair. So, Michael Crichton could describe the laboratory work in convincing detail in his book, and it was an equally believable part of the film. But could it really work?

At one level, Crichton was hugely prescient. He had early on trained as a medical doctor, and so was comfortable with the medical and biological literature. He was quick to appreciate the potential of the new cloning method, the polymerase chain reaction (PCR), which had been developed in 1983 by Kary Mullis, and for which he received the Nobel Prize for Chemistry a decade later. PCR allows medical doctors and biologists to amplify a single copy or a few copies of a segment of DNA to thousands or millions of copies. Before PCR, large, purified samples were needed before any assay could be carried out, and this made molecular biology and genetic engineering hugely costly and time-consuming. After PCR came the genetic revolution, with all its economic consequences for the future of medicine and agriculture.

Here is a step-by-step outline of how to clone a dinosaur, or at least in the way they did it in *Jurassic Park*:

1. Extract blood from a mosquito in amber by inserting a fine-needled syringe.
2. Concentrate the DNA by spinning the blood sample very fast in a centrifuge.
3. Take a small sample of the concentrated DNA and clone (= multiply) it.
4. To clone the DNA, it is cut into sections; these are inserted into bacteria that then copy the segments, and divide many times. So, one copy becomes many copies.
5. The multiplied DNA sample is then injected into the egg of a modern frog (the frog DNA has already been removed, and the egg is just a single cell at the start).
6. The dinosaur DNA takes over the working of the frog's egg and, instead of having the genetic code for a frog, it has the genetic code for a dinosaur.
7. The scientists then wait for the frog's egg to develop – into a dinosaur.
8. The egg does not turn into a tadpole, which would then turn into a frog. The genetic code has been replaced, and the single-celled egg then divides into two, four, eight, sixteen... but each of those cells is being guided by the dinosaur DNA to be a dinosaur cell.
9. The outside cells form a hard eggshell, and so the egg looks like a dinosaur egg, with a hard shell like a bird's egg, not a soft and squishy frog's egg.
10. Then comes the day of hatching, which is shown in the film. A crack appears in the hard, white shell, a scaly snout pokes out, and then a head, and at last a small dinosaur hops out, ready to snap and bite and looking for food.

So, this seems quite straightforward. Molecular biology and genetics have advanced so much in the century since Conan Doyle wrote *The Lost World* that anything seems possible. Can we really hope to use modern molecular techniques to bring ancient animals back to life?

Has dinosaur DNA ever been identified?

I remember reading Crichton's book when it came out in 1990, and palaeontologists were intrigued. Many, I'm sure, read the book hoping to pick holes in it, but mostly they had to admit the scenario was plausible. The technical problem of implanting the DNA and making a dinosaur was surely very tricky, but then it was not universally rejected that we might some day recover real dinosaur DNA. And, indeed, that's just what happened, even before the film came out in 1993.

In 1992, Raúl Cano and his West Coast colleagues caused a sensation. They announced that they had extracted DNA from a fossil bee, preserved in amber from the Dominican Republic in the Caribbean, dated at up to 40 million years old. This was not from a dinosaur, but any record of ancient DNA was a start. A year later, Cano and colleagues revealed that they had extracted DNA from a plant in the Dominican amber, and something even more amazing: DNA from a weevil preserved in amber from the Lebanon, dated at 120–135 million years old.

Extracting organic molecules from fossils in amber was all the rage around 1990, and independent labs announced their extractions of DNA from a termite in the Dominican amber and from a beetle in Lebanese amber, apparently confirming the Cano team's results. These reports of DNA from an ancient weevil and an ancient termite, insects that lived at the same time as the dinosaurs, could not have been better timed. Admittedly, the teams had not extracted the blood of a dinosaur from a mosquito, but they had apparently proved that DNA could exist for millions of years, from the age of the dinosaurs. So maybe Crichton's imaginative story could come true.

Then, in 1994, came the bombshell. Dinosaur DNA was announced, in a paper in *Science*, the leading US science journal. The discovery was reported by *New Scientist* at the time:

> Dinosaur bones retrieved from a coal mine have given up some of their secrets to scientists at Brigham Young University in Utah. Scott Woodward and his team have extracted short stretches of the dinosaur's DNA, although they have a long way to go before they can reconstruct a whole creature as in Michael Crichton's *Jurassic Park*.

Woodward and his team extracted DNA from nine samples during a year of experiments, but the success rate was only 1.8 per cent. 'If we hadn't

A mosquito preserved exquisitely in amber.

gotten one in an early round (of experiments), we probably would have given up,' he admitted. As *New Scientist* continued, 'The DNA came from two unfossilised pieces of bone from deposits 80 million years old in a Utah coal mine. Although the prehistoric owner of the bones has not been identified, their size and location makes Woodward "confident they are dinosaur bones".'

However, within a year, the story had been blown. The 'dinosaur DNA' was in fact human. Woodward denied this initially, and promised follow-up work, but his paper was one of several under close scrutiny at the time. In most of these early experiments, the authors had not taken sufficient precautions to avoid the risk of contamination. A key property of PCR is that it clones multiple copies of DNA from very small samples, and indeed all it can take is a drop of sweat or a sneeze by one of the technicians in the lab, and the whole study is spoiled.

Do organic molecules survive in the fossil record?

The risk of contamination when measuring ancient molecules in fossils, especially ancient DNA, was highlighted from the 1990s. Critics noted the risk of contamination, not just by human DNA, but also by DNA from modern animals. Indeed, the risk in the early reports of plant and insect DNA was that the fossil samples were being analysed in labs that also processed DNA of modern relatives. So, for example, the DNA extracted from the Cretaceous weevil or termite in amber could easily be mixed up with DNA from modern weevils and termites. Something stricter was needed for all future studies of ancient DNA.

Since the 1990s, the technology of ancient DNA labs has advanced enormously to exclude all risk of contamination. The strict measures include the following: (1) everyone entering the lab has to remove their outer clothing and change into a cleaned suit, which has a hood to cover the hair, and a face mask to prevent the technicians from breathing or shedding hair onto the samples; (2) all ancient DNA is studied in one lab, and modern DNA is studied elsewhere, to avoid any risk of mixing; (3) every analysis is repeated in another lab, to ensure any contamination is highlighted; and (4) the ancient DNA lab is sterilized every night by bathing everything in ultraviolet rays overnight, so any living organisms, whether flies or bacteria, are killed.

These precautions should ensure that contamination is excluded, but how long can DNA survive? One of the most prominent critics of the reports of ancient DNA was biochemist Tomas Lindahl in London. He noted that DNA degrades in days, months, or years. So, under normal conditions, there wouldn't be much DNA left after 100 years, let alone 100 million years. Subsequent studies have shown that it is possible to extract DNA from museum skins and skeletons of recently extinct animals, such as 100-year-old specimens of the horse-like quagga from southern Africa, and the dodo, which died out before 1681. Records were pushed back even further to ancient Egyptian mummies from 5,000 years ago, then 10,000-year-old mammoths, and finally, in 2013, DNA was retrieved from a horse dated as 700,000 years old.

This horse example is much older than all other examples, and the DNA is fragmented into many short sequences. Indeed, even after 100 years, the quagga DNA had broken down substantially, and this makes its interpretation very tricky. Once the fragments consist of fewer than ten base pairs, it might seem impossible to reconstruct anything like an original DNA sequence of any length; the only solution is to use massive computing resources to crunch through every possible combination of answers until something plausible emerges. So, it's unlikely that any dinosaur DNA, or indeed any DNA older than 1 million years, let alone 100 million years, will ever be retrieved.

It has become obvious, after all this effort, that DNA is not a tough molecule. Indeed, chemists identify a spectrum of the toughness of organic molecules, a scale from those that can stand a great deal of pressure and heat through to those that break down at the slightest challenge. During fossilization, most biological tissues break down soon after death; they decay in air or soil or water, and animals may scavenge the flesh, and bacteria break it down further. Only in rare cases is a skeleton not quickly stripped of skin, muscle, and internal organs, and this usually involves covering by water and sediment and the exclusion of oxygen. In such cases, the biomolecules may be buried, but then they have to survive high pressures and temperatures, and most will disappear, or become so modified that they cannot be recognized.

Molecules that survive include lignin, chitin, and melanin. Lignin is a structural molecule that makes up the wood inside trees, and it can survive for hundreds of millions of years. So too can chitin, a carbohydrate that makes up the tough cuticle of arthropods – think of the hard wing casings of a beetle. Finally, as we have seen in Chapter 4, melanin is a pigment that generally gives a black or dark brown colour, and it is found in feathers and hairs (and in dark skin and freckles), as well as in the retina of the eye, the ink of squid, around the liver and spleen, and in the membranes of the brain. It's thanks to these refractory properties of melanin that the colour of feathers of fossil birds and of dinosaurs could be determined, as we have seen.

Experiments on chitin and melanin have shown how the complex molecular structure changes as pressure and temperature are increased (see overleaf), and indeed the ancient fossil forms show the same detailed chemical characteristics of these experimentally produced samples. As expected, when similar experiments were done with biomolecules such as DNA, it simply broke up completely and nothing recognizable survived the fossilization experiment. Therefore, any organic chemist could have said that if you want to find ancient organic molecules, look for examples of lignin, chitin, and melanin, and not ancient DNA. Yet, there have been persistent reports of the long-term survival of dinosaur blood. In a delicious twist, work published just as this book was going to press showed that my scepticism was partly right, but also partly wrong.

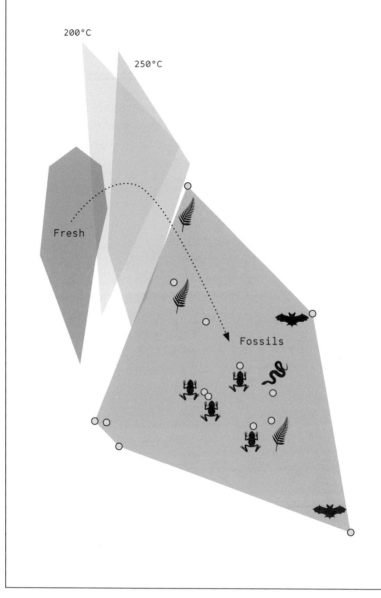

200°C

250°C

Fresh

Fossils

PC2 (15.686% variance)

PC1 (23.436% variance)

Experiments show how melanin decays, very
slowly, under high temperature and pressure; the
experimental values are close to those in the fossils.

Can we identify dinosaur soft tissues and blood?

It was a huge disappointment to realize that DNA can't survive for more than a few thousand years. So, all those reports of DNA from insects, plants, and bacteria that were millions of years old have been rejected. But what if other kinds of proteins might survive in dinosaur fossils, perhaps specific proteins within the bones? Fresh hope came in 1997 with the report of traces of dinosaur blood. A team of researchers from Montana State University, led by Mary Schweitzer, reported that they had extracted proteins and blood compounds from unusually well-preserved bones of *Tyrannosaurus rex*. If this were true, it would bring us close to the physiology of the dinosaurs – the structure of their haemoglobin might give clues about its oxygen-carrying abilities, and hence whether dinosaurs were warm-blooded or not.

Mary Schweitzer was inspired in her quest for ancient proteins by an exceptionally well-preserved tyrannosaur skeleton. 'In parts it was almost the same as modern bone, with no mineral infilling,' she said. A dense outer layer of bone seemed to have stopped water washing in, and so the interior bone was apparently as good as new. Schweitzer identified proteins and possible DNA in these inner zones. She reported the buzz of excitement at the time:

> The lab filled with murmurs of amazement, for I had focused on something inside the vessels that none of us had ever noticed before: tiny round objects, translucent red with a dark center. Then a colleague took one look at them and shouted, 'You've got red blood cells. You've got red blood cells!' It was exactly like looking at a slice of modern bone. But, of course, I couldn't believe it. I said to the lab technician: 'The bones, after all, are 65 million years old. How could blood cells survive that long?'

The bones with possible blood cells were tested. The bones indeed seemed to contain haem, the oxygen-carrying part of the haemoglobin molecule of the blood. Haem is red, giving blood its red colour, because it is rich in iron, and the red colour appears when it combines with oxygen, a little like the colour change of iron when it rusts. Many other scientists queried these reports, however, and suggested that the iron-rich traces in the bone were nothing to do with blood or blood products, and might just have been iron minerals that grew in the bones long after their burial.

After much criticism, some fair and some probably unfair, Mary Schweitzer and her team published a follow-up paper in *Science* in 2005, entitled 'Soft-tissue vessels and cellular preservation in *Tyrannosaurus rex*'. Her team dissolved away the calcium phosphate of the hard portions of some limb bones, and were left with a residue consisting of narrow vessels which contained round bodies that could be squeezed out. The demineralized bone matrix was fibrous and retained some of its original elasticity – pretty amazing for a 70-million-year-old fossil. In a later study of the same materials, Schweitzer and her colleagues carried out a battery of biochemical tests to show that the elastic fibrous strands were composed of collagen, as in the original bones.

Bones are typically composed of two main materials: mineralized needles of apatite, a form of calcium phosphate, which are embedded in the fibrous protein collagen. It's this combination of elastic protein and hard mineral that gives live bones their interesting properties of being able to bend (to some extent) and then to snap in a brittle manner. Where the apatite crystals are absent, the collagen forms cartilage, the flexible material that stiffens our ears and nose, and that forms the bulk of the skeleton of a shark.

Soon after, in 2008, Thomas Kaye and colleagues reinterpreted all these fossils as artefacts. They said that the supposed blood vessels were probably bacterial films, and the possible red blood cells were crystals of pyrite, the mineral form of iron sulphide. Mary Schweitzer rejected these criticisms, and her work was seemingly confirmed by further reports by another team in 2015 of collagen and red blood cells from eight Cretaceous dinosaur bones.

However, in a paper published in 2017, Manchester-based Michael Buckley and colleagues showed that the *T. rex* collagen comprised mainly laboratory contaminants, soil bacteria, and bird-like haemoglobin and collagen. In particular, they found that the supposed dinosaur proteins matched modern ostrich sequences, an easy mistake to make when such modern samples might have been in the same laboratory that worked on the fossil materials. Then came some clarity. In a 2018 paper, Jasmina Wiemann, a PhD student at Yale, led a group that looked again at the blood vessels and other pieces of brownish material left after fossil bones had been processed to remove all mineral components. She applied sophisticated tests and found that the vessels and tissues were real, but not made of original proteins, except perhaps collagen. The others had decayed to alternate forms called N-heterocyclic polymers – so in fact Mary Schweitzer was right that she had found blood vessels, skin cells, and portions of nerve endings, but their proteins had been substantially converted during fossilization.

Collagen might also be preserved, although care must be taken to be sure it is original and not contaminated. Another bone protein, osteocalcin, was reported in 1992 from the bones of two Cretaceous dinosaurs by the Dutch researcher Gerard Muyzer. Osteocalcin is found in the bones of all vertebrates, and it functions like a hormone in stimulating the repair of bones, as well as other physiological functions. Osteocalcin is a tough protein that is bound into the bone minerals very securely, and it's that relationship that seems to protect it from decay. It is also a relatively small protein, consisting of about fifty component amino acids. Complete osteocalcin molecules from a 55,000-year-old fossil bison were sequenced in 2002. Maybe we can hope that dinosaur osteocalcin will be sequenced some day.

Can we identify the sex of a dinosaur?

Palaeobiologists have long suggested that some dinosaurs, at least, were sexually dimorphic, meaning that males and females had a different appearance, as we saw in Chapter 4. In the early days, this was suggested for the horned ceratopsians and the crested hadrosaurs. In the Late Cretaceous, these were the dominant herbivores, and in each case, the skeleton is nearly identical among all the species, but the headgear differs. In one famous case, it later turned out that all the males lived in one spot at one time, and all the females, with some differences in their skulls, happened to live at another place and at another time. Collapse of hypothesis!

Sexual dimorphism in dinosaurs has come back to the fore, however, now that we can identify in some detail the colours and patterns of their plumage. It is now accepted that the function of the feathers of many dinosaurs was probably display, and the stripes and crests suggest pre-mating displays by males, as among birds, and hence a key role for sexual selection in the evolution of dinosaurs, as we saw in Chapter 4.

Amazingly, it is possible to sex some dinosaurs based on unequivocal evidence. Most female birds show a specialized kind of bone called medullary bone, which is spongy bone that fills up the medullary cavity – the core – of certain limb bones. In modern birds, it was first noted in 1934 in pigeons, and then observed in sparrows, ducks, and chickens. The medullary bone can be laid down very quickly, and recycled very quickly, acting as a store for calcium, which can be mobilized rapidly when needed to form an eggshell. Later observations have shown that this

The most amazing fossil:
two *Confuciusornis* on
one slab, one female
and one male (with
long, banner-like tail
feathers).

Chapter 5

occurs in all modern birds. Physiological experiments show that it builds up in the core of many bones throughout the skeleton just as the female bird begins to lay down yolk, and then it diminishes as calcium passes into the developing eggshell. Medullary bone development and transfer are cyclical according to the seasons of the year, and they are controlled by oestrogen and other hormones relating to the breeding cycle.

Medullary bone was first noted outside modern birds in *Tyrannosaurus* by Mary Schweitzer in 2005. Since then, it has been reported from other theropod dinosaurs, the ornithischians **Tenontosaurus** (see overleaf) and *Dysalotosaurus*, and the extinct birds **Confuciusornis** (see overleaf) and *Pinguinis*. The report on *Confuciusornis* by Anusuya Chinsamy-Turan and colleagues from the South African Museum in Cape Town was especially telling, as it proved that the medullary bone identified in the fossils occurred in female specimens (see pl. xiii). Among the thousands of specimens of the crow-sized *Confuciusornis* in Chinese museums, two sexual forms had been identified. One classic specimen shows a male and a female bird on the same slab – the supposed male has single long, banner-like tail feathers, while the supposed female does not. So, as with modern birds, the male has the ridiculous adornments, to show off to the more sensible, drab female what a tough character he is, and so what a good father he will make. Chinsamy-Turan and colleagues identified medullary bone in the inner cavity of a microscopic thin section, spongy bone tissue quite distinct from the regular, more compact bone. The medullary bone was only ever found in females, never in males – although not in all females, since they were not all in reproductive mode when they were killed.

Other cases, though, have been disputed, including the reports of medullary bone in larger dinosaurs, including *Tyrannosaurus* and *Allosaurus*. Another explanation of the spongy bone in these large dinosaurs is that it might have been associated with growth spurts. Some of the larger dinosaurs grew really quite fast, and put on hundreds of kilograms of weight in a few months, as we shall see in Chapter 6, and so they needed to capture and mobilize calcium quickly for that purpose. There is no doubt about the reproductive purpose of medullary bone in living birds, and probably also in fossil birds, but maybe it was found only in smaller dinosaurs for which egg-laying was a big effort, as in birds today.

Delving into dinosaur bones to understand their physiology and sexual habits is one thing. How about the opening theme of this chapter – could we ever engineer a living dinosaur?

Genus: **Tenontosaurus**

Species: *tilletti*

Genus: **Confuciusornis**

Species: *sanctus*

Named by:	Lianhai Hou and colleagues, 1995
Age:	Early Cretaceous, 125 million years ago
Fossil location:	China
Classification:	Dinosauria: Saurischia: Theropoda: Maniraptora: Avialae (birds)
Length:	0.5 m (1¾ ft)
Weight:	0.5 kg (1 lbs 2 oz)
Little-known fact:	This is the best-known fossil bird of all time, with thousands of specimens in museums.

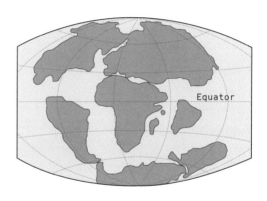

Named by:	**John Ostrom, 1970**
Age:	**Early Cretaceous, 115–108 million years ago**
Fossil location:	**United States**
Classification:	**Dinosauria: Ornithischia: Ornithopoda: Iguanodontia**
Length:	**6.5–8 m (21–26 ft)**
Weight:	**0.8–1 tonne (1,764–2,205 lbs)**
Little-known fact:	**The first fossils were found in 1903, but they were not fully understood until complete skeletons were excavated in the 1960s.**

Could we bring dinosaurs back to life by genetic engineering?

Perhaps we will never recover any dinosaur DNA, because this biomolecule is well known to decay rapidly. But what about cloning? We've all heard about Dolly the cloned sheep, and there are constant suggestions that scientists could apply such techniques to bring mammoths back to life. Could this work?

Dolly the sheep became a scientific sensation when her birth was announced in 1997. In 1995, a group of scientists at the Roslin Institute, near Edinburgh, were looking for a way to genetically modify farm animals. They cloned two sheep, Megan and Morag, from embryo cells grown for several weeks in the laboratory, but Megan and Morag did not develop very far and they could not be brought to birth. Dolly was born on 5 July 1996, although the world didn't find out about her until early the following year. She was the first mammal cloned from an adult, rather than embryonic, cell, and her birth – splashed over newspaper front pages – brought the issues surrounding cloning to breakfast tables around the world.

Sadly, Dolly died in February 2003. Did she die young because she was a clone? Was Dolly unnatural, a Frankenstein monster created by crazy scientists in the laboratory? There are many debates about the ethics of cloning. Some people object to the whole idea of genetic engineering, for either religious or political reasons, while others say that scientists should be free to carry out experiments that push back the frontiers of human knowledge. In food production, genetic engineering of plants has been a regular part of agriculture now for decades, and most sweetcorn, and many other grains and pulses you consume, have been genetically engineered to improve the crop yield or nutritional content.

Cloning means literally 'making a copy', and the idea of cloning is to find a way to make an egg develop into an adult plant or animal, using its DNA, but without the normal process of the male fertilizing the egg in the female. The steps in cloning in the laboratory are: (1) remove some complete DNA from the cells of the animal or plant you want to clone; (2) remove the nucleus from an egg of the host animal; (3) inject the DNA into the emptied host egg; and (4) grow this cell in the womb of a living mother animal, probably a very close relative. This is how Dolly the sheep was cloned, and nurtured through to adulthood.

As a first step towards cloning an extinct species, biotechnologists have focused on cloning at-risk species. One example is the gaur, a large wild ox species that lives in India and southeast Asia. It is huge, about 2 metres (6½ feet) tall at the shoulder, and weighs not much less than a tonne (2,205 pounds). It used to be common, but its population size has been reduced to some 36,000 by hunting. So, a biotech company, Advanced Cell Technology (ACT) in Massachusetts, USA, decided to try the experiment of cloning a gaur. They wanted to do something a little different from the case of Dolly the sheep, and to use another species to act as mother. Although they could have transplanted the gaur eggs into a female gaur, they wanted to use this as a test of concept for bringing extinct species back to life. If a species is extinct, there is no living mother, and a female from a related species must be used.

In 2001, the scientists announced the successful birth – and death – of the first endangered animal clone. The baby bull gaur, Noah, died within forty-eight hours. Researchers from ACT said the problem was unlikely to be related to the cloning procedure itself. The clone had been carried by a domestic cow called Bessie. Noah was produced in a cross-species cloning procedure. The genetic material taken from the skin cells of a male gaur that had died eight years previously was fused with the emptied egg cells of common cows. From a total of 692 eggs used in the experiment, only one live clone was produced – Noah. Noah died of a common illness (dysentery) that probably had nothing to do with the fact he was a clone. Bessie, the surrogate, remained fit and well.

The first attempt to bring an extinct species back to life was with the Pyrenean ibex. This mountain goat lived in the Pyrenees until it was hunted to extinction. The last Pyrenean ibex alive was a female named Celia, and she was found dead in the year 2000. Before she died, Spanish biologists captured her and took a tissue sample from her ear. ACT, the cloning company that had produced Noah, then announced that the Spanish government had commissioned them to clone the Pyrenean ibex, bringing it back from extinction.

The company used the sample from Celia's ear, and they took adult body cells and fused them with egg cells from goats that had had their nuclei removed. The ibex–goat embryos were then transferred into domestic goats, which acted as surrogate mothers. In 2009, after several years of failed attempts, the first two Pyrenean ibex kids were born to their goat mothers – but sadly they died soon after birth. Further cloning attempts were made in 2014, and if future cloned young survive, then an extinct species will have been resurrected.

There are many other cloning projects working with the genetic material of recently extinct mammals, a new science called de-extinction

The extinct Pyrenean ibex, drawn when the
species had not yet died out.

1. Preserved cells
extracted from frozen
mammoth carcass

4. Mammoth genes
are placed inside
elephant egg cells

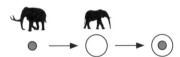

2. Nuclei extracted
from mammoth cell

5. Egg is placed
in elephant womb

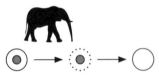

3. Egg cells taken
from elephant, DNA
material is removed

6. Elephant gives
birth to live mammoth,
effectively a clone
of the frozen carcass
that yielded undamaged
cells

How to clone a mammoth.

or resurrection science. Targets include not only the Pyrenean ibex, but also the thylacine or Tasmanian tiger, the aurochs, a European wild cow, the quagga, an extinct species of zebra, and the passenger pigeon. More daringly, several teams are talking about cloning and de-extincting the mammoth, using a female Asian elephant as the surrogate mother. So far, though, it's all talk and no results.

If it is so hard to succeed with multi-species surrogacy when there are really close living relatives (such as the Pyrenean ibex and goat), how much more technically difficult it is going to be to cross species boundaries from Asian elephant to mammoth? And, as for dinosaurs, which species would serve best as their living surrogate? Of course, they laid eggs, like birds and crocodiles, so the mother would not have to carry the bizarre dinosaur embryo inside, but, as in *Jurassic Park*, the biotechnologists would have to engineer the dinosaur embryo and induce it to develop inside the egg of a living species. We're a very long way away from seeing this happen.

Can we say anything about the dinosaurian genome?

The genome is all the genetic code contained in our cells. Molecular biologists talk about two aspects of the genome – its overall size (in other words, the total number of genes, those portions of the genome that have specific functions) and how it is organized in the chromosomes, the X- and Y-shaped structures inside the nucleus.

In terms of overall genome size, it seems that birds, theropods, and sauropodomorphs had small genomes, whereas ornithischian dinosaurs had much larger genomes. Nobody has ever seen the genome of a dinosaur, but the size of the genome has been shown to relate to the average size of cells. By measuring the size of cells within fossil bones, Chris Organ and colleagues were able to speculate about overall genome size. They suggested that the small genome size of those dinosaurs, and birds, was related to their warm-bloodedness, and especially the origin of flight in theropods and birds.

The chromosome organization in dinosaurs has also been reconstructed according to a 2018 paper. In this paper, molecular biologist Rebecca O'Connor and her team from the University of Kent mapped the DNA from different chromosomes of modern birds and reptiles, and looked for shared components. By comparing the complete genetic sequence of birds and turtles, they could be sure they were including all relatives of turtles and birds, and that includes dinosaurs.

The team used fluorescent labels, or 'DNA probes', to identify shared portions of the genetic sequences between turtles and birds, and then they could assume these had been present in the common ancestor of both turtles and birds – which would have been a reptile that lived over 300 million years ago, and well before the dinosaurs originated. But those shared features of the DNA of turtles and birds were almost certainly present in dinosaurs.

They concluded that most elements of the modern bird genetic code and its arrangement in 40 chromosome pairs were present in the ancestral form, and so this reorganization had occurred before the origin of dinosaurs, and likely was shared with them all. This is a remarkable discovery because it implies that some special features of the genetic code of birds had arisen much earlier than expected. For example, birds have the high number of 40 chromosome pairs (compared to 33 pairs in turtles and 23 pairs in humans), and the new evidence is that the multiplication of chromosomes happened early, and that dinosaurs all shared that change in genetic architecture.

Identifying ancestral gene sequences shared by all reptiles and birds provides a template of the minimal composition of the dinosaurian genome. The authors wisely say nothing, however, about whether this could ever be a basis for cloning dinosaurs. They do go so far as to say 'that the overall genome organisation and evolution of dinosaur chromosomes...might have been a major contributing factor to their morphological disparity, physiology, high rates of morphological change and ultimate survival'. They also note that the apparently very early acquisition of bird-like chromosomes might correlate with the discovery of the early acquisition of many supposedly unique bird characters (including feathers, hollow bones, and the wishbone) by theropod dinosaurs.

The news about bringing dinosaurs back to life is not promising. The methods can all be identified, but DNA does not survive for long, and so there is currently no prospect of obtaining any dinosaurian DNA. Without the genetic code in the DNA, the whole *Jurassic Park* scenario collapses. Even if we could clone extinct animals – which we have so far failed to do – we still need that genetic code.

The pursuit, however, has not been fruitless. Some have perhaps allowed their enthusiasm to outpace the quality of their data in making announcements of DNA, blood cells, and other miraculous examples of preservation from the age of the dinosaurs, but these are tricky areas at present. Others are equally damning of our case for melanin and melanosomes. The glory of these fields of palaeobiology is that they are perfectly interdisciplinary, drawing in the expertise of molecular biologists, geneticists, and organic chemists. Long may the hunt for extraordinary fossils continue!

Chapter 6

From Baby to Giant

Dinosaurs were often huge – which is a conundrum in itself – but they started life from relatively small eggs, so they either had to grow up very fast or live for a very long time. The subjects of growth and size go to the heart of some key questions about dinosaurian palaeobiology, and they are topics that have given rise to quite wild speculation over the years.

Even if the dream of *Jurassic Park* may never be realized, we still know a great deal about the development of dinosaurs from egg to adult. Skeletons of dinosaurs show all growth stages, from embryos in the egg, through hatchlings and juveniles, to adults. This is what drew palaeontologist Greg Erickson, now a professor at Florida State University, into the subject. In his words:

> When I began we didn't know much about the basic biology
> of dinosaurs, such as how long they lived, how fast they grew,
> aspects of their physiology, reproduction, etc. My career goal
> was to develop methods to glean such information. I began by
> figuring out how fast they grew, a proxy for metabolic rates.
> At the time it was debated whether they grew slowly like
> cold-blooded reptiles, perhaps taking over a hundred years to
> mature, which frankly seemed unlikely to me, or much more
> rapidly like warm-blooded birds and mammals.

Although Erickson liked palaeontology and geology as a kid, he did not entertain a career in those sciences when he first entered college at the University of Washington in Seattle. In fact, he graduated still unsure of what he wanted to do. He ended up getting a degree in geology, and a palaeontologist invited him to participate in some expeditions, and encouraged him to pursue an academic career. After graduating he worked in a miserable job as a construction worker, which provided ample time for him to mull over what the professor had said. 'Learning what I didn't want to do showed me the way to what I did want to do – palaeontology.'

As a young professor, he recalls, 'I was at the Field Museum in Chicago and noticed growth lines in the bones of Sue, the largest known *T. rex*, which had been purchased for 8.36 million dollars. This led me to ask, "Can I cut up your seemingly priceless dinosaur?" Fortunately, after much debate among the museum higher-ups, I got the approval, and that springboarded the work.' Growth rings in the bone are key to understanding the age of a dinosaur skeleton, and that enabled Erickson and others to draw up growth curves for dinosaurs, as we shall see.

Dinosaurs all hatched from eggs. Birds and crocodiles lay eggs with hard shells made from calcium carbonate, and it is no surprise that dinosaurs laid eggs as well. Fossil dinosaur eggs were first reported, not from North America or Mongolia, but from the Cretaceous of the south of France in 1859.

Jean-Jacques Pouech, a Roman Catholic priest, and head of Pamiers Seminary, was exploring the rocks in the foothills of the Pyrenees, and he found great plated fragments of shell covered with pustular, or regularly roughened, surfaces. He reported:

The egg of the sauropod *Hypselosaurus* found in France; the eggshell is shattered but repaired.

The most remarkable are eggshell fragments of very great dimensions. At first, I thought that they could be integumentary plates of reptiles, but their constant thickness between two perfectly parallel surfaces, their fibrous structure, normal to the surfaces, and especially their regular curvature, definitely suggest that they are enormous eggshells, at least four times the volume of ostrich eggs.

Pouech identified the French fossil eggs as those of giant birds.

Much more famous were the discoveries of dinosaur eggs and nests from the Cretaceous of Mongolia, in the 1920s. Roy Chapman Andrews, the explorer hired by the American Museum of Natural History (AMNH), led several large-scale expeditions to Mongolia, working from his base in war-torn Beijing, and driving northwards towards Ulaanbaatar in a cavalcade of black model-T Fords, and then off into the remotest deserts with hundreds of gallons of water, stacks of food, and rifles. His expeditions were really about going into the unknown, as Beijing was at that time over-run by warlords struggling to take power in China. This was the far-from-peaceful base that Chapman Andrews used to prepare equipment and supplies before driving 1,000 kilometres (620 miles) into the Gobi Desert. Despite the risks, on the first expedition, the team excavated dozens of dinosaur skeletons, as well as nests.

The most famous exhibit at the AMNH was of several small ceratopsian (horn-faced) dinosaurs, **Protoceratops**, clustered round their nests, and with the predatory dinosaur *Oviraptor*, which means 'egg thief', scouting around and threatening them. The nests were about 1 metre (3 feet) across and each contained twenty to twenty-five eggs, arranged in concentric circles. The eggs are long and cylindrical, and they were laid with their pointed end inwards, so making it easy for them to roll naturally into a circular array. When the first specimens were put on show in New York they attracted crowds, and the public loved the narrative of the humble herbivore *Protoceratops* trying to defend its nest from the wicked egg thief. But was this true?

Dinosaurs started so small, and yet some grew to be truly huge. This poses a number of interesting conundrums. By scaling to their adult size, dinosaur eggs should have been much larger than they were; and of course the largest dinosaurs, ten times the size of an elephant, just defy understanding – how could they be so huge when nothing on land today even approaches their dimensions? New studies suggest how dinosaurs achieved the impossible.

Genus:	**_Protoceratops_**
Species:	_andrewsi_

Named by:	**Walter Granger and William Gregory, 1923**
Age:	**Late Cretaceous, 84–72 million years ago**
Fossil location:	**Mongolia**
Classification:	**Dinosauria: Ornithischia: Ceratopsia:** **Protoceratopsidae**
Length:	**1.8 m (6 ft)**
Weight:	**83 kg (183 lbs)**
Little-known fact:	**Skulls of _Protoceratops_ might be the first fossils ever seen, when ancient Greeks wandering over the Gobi Desert thought their skulls were the remains of dragons.**

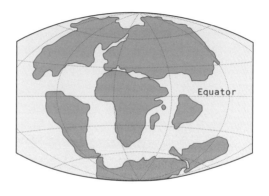

Why were dinosaur eggs and babies so small?

Were dinosaur eggs and babies unusually small? Certainly, by scaling to modern birds, the largest dinosaur eggs should have been about the size of a Smart car, maybe 2 metres (6½ feet) long, and yet the largest dinosaur eggs were only 60 centimetres (23½ inches) long and about 20 centimetres (8 inches) in diameter, so quite long and sausage-shaped. Even these eggs, reported from China in 2017, were comparatively large by dinosaur standards; dinosaur eggs rarely exceed the size of a rugby ball or American football, about 30 centimetres long. These giant eggs from China contained tiny bones, which identified the embryo inside as a relative of *Oviraptor*, the so-called 'egg thief' of Mongolia. These were from a much larger relative, maybe 2 tonnes (4,410 pounds) in weight when fully grown, but only a couple of kilograms at birth.

The scaling is all wrong when you compare these dinosaurian examples with birds. The relationship isn't exactly simple – it's an exponential curve, and the proportion of egg mass to adult female body mass also changes. Small birds such as hummingbirds and tits have relatively enormous eggs, making up 20 per cent of the female body mass, whereas in larger birds such as gulls and ostriches, the proportion is 5 per cent or less. Even allowing for continuing relative decline in the proportion as adult body size increases, a 10-tonne dinosaur should lay an egg of, say, 2 per cent of adult body mass, so 200 kg; and a 50-tonne sauropod might have produced an egg of 500 kg (say 1 per cent of 50,000 kg). So, their eggs, perhaps weighing at most 2–3 kg (5–7 pounds), were far too small. But why?

It's a combination of basic mechanics and energy saving. In terms of mechanics, the thickness of an eggshell is proportional to the volume of the egg – after all, the mineralized shell must be robust enough to prevent the egg from collapsing. A hen's eggshell is a fraction of a millimetre thick, whereas an ostrich eggshell is 2–3 millimetres thick, and a hypothetical 500-kilogram (1,100-pound) dinosaur egg would have to have an eggshell several centimetres thick. This would be catastrophic for two reasons. Oxygen could not percolate in, nor carbon dioxide out, through such a thick crystalline structure, so the embryo would die; but also, when it came to hatching, the poor baby, as large as a pony, would struggle to bash its way out.

The energy-saving aspect of laying small eggs is part of the overall life strategy of dinosaurs. Ecologists often characterize animals according to whether they emphasize quantity or quality of their offspring. Those that focus on quantity simply produce as many eggs as they can, but do not invest much effort in them. A classic example is the cod fish, which produces more than a million eggs at a single sitting, seemingly an excellent way to fill the world's oceans with cod. In fact, the eggs and babies are useful food for many predators, and only two or three survive to adulthood, but that is enough to keep the cod species alive (barring over-fishing). The quantity-focus life strategy seems wasteful in resources – all those eggs and babies are produced, but 99.9999 per cent end up as food for others.

By contrast, mammals, including humans, invest heavily in the care of their young, the quality strategy, and so they tend to produce fewer young at a time, and seek to ensure that they survive. However, this strategy is also wasteful in resources because the parents, or just the mother, devote a large portion of their lives to child care rather than their own survival.

Dinosaurs did not produce a million eggs at a time, but more typically three to five for some species, and fifteen to twenty for others, such as *Oviraptor*. This compares with modern birds, which have clutch sizes ranging from a single egg to eighteen, but with an average of three, as shown in a recent analysis of more than 5,000 species of living birds. The single-egg birds are mainly large sea birds such as albatrosses, shearwaters, and petrels, that struggle to feed their chick, and so could not rear more than one or two simultaneously. Large clutches, typically ranging from seven to eighteen eggs, are produced by temperate-climate species such as ducks, pheasants, and partridges, and they rear their young at specific times of the year when seasonal food supplies are rich.

So, which of these two parenting strategies did dinosaurs adopt? On the one hand, crocodiles and other living reptiles mainly adopt the quantity life strategy – lots of eggs that they abandon after laying. Birds, on the other hand, are quality strategists, like mammals, laying usually modest numbers of eggs and caring for the young after they hatch. Dinosaurs seemingly did a bit of both, but tended towards the quantity strategy, laying a reasonable number of eggs, and then abandoning them to their fates. Importantly, the small size of the eggs contributed hugely to saving energy devoted to reproduction. This retention of reptilian behaviour is a major plank in theories for how dinosaurs could reach such large sizes, as we shall see.

What do we know about dinosaur embryos?

When palaeontologists first began collecting dinosaur eggs, the specimens were often simply broken pieces of eggshell. Even when palaeontologists found complete eggs, they did not think to look inside, even though the fact the egg was complete indicated it had not hatched, and so might very well contain an embryo (unless, of course, like our breakfast eggs, it had never been fertilized).

Later, some damaged dinosaur eggs showed hints of tiny bones inside, but these had to be exposed by cleaning away sandstone by laborious efforts under the microscope. But this careful work with a needle could still cause damage to the tiny, delicate bones, and it seemed the study of embryo skeletons was doomed.

Scanning has changed all that. As we saw earlier, CT scanners can produce remarkably detailed information about even tiny fossils buried in the rock, but not every lab has a CT scanner. A combination of classical preparation methods and CT scanning has revealed a lot about one group of dinosaur embryos. In 1976, James Kitching, world-famous fossil collector in South Africa, excavated a clutch of six dinosaur eggs and brought them back to the Bernard Price Palaeontological Institute in Johannesberg. There they sat for some time, before an international team began the research.

Diane Scott, fossil preparator at the University of Toronto in Canada, did some delicate work with the needle to clear the rock, grain by grain, from one of the South African embryos. The foetus was confirmed as the young of **Massospondylus** (see overleaf), the most abundant plant-eater of its day, and 5 metres (16½ feet) long as an adult. The embryo is curled up inside the egg, with the head and body clearly visible, arms and legs tucked neatly below, and the tail curling round the back. The bones of the skull are disarticulated, which is not unexpected because bones are not fused together in such young individuals. In fact, this is true of all babies, and parents know that their children have a fontanelle, or gap, between the frontal and parietal bones on top of the head when they are very young. In the case of the South African *Massospondylus* embryo, the dinosaur head is 1 centimetre long, making the baby only 15 centimetres (6 inches) long in all, representing 3 per cent of adult length at birth. By contrast, human babies are 25–30 per cent of adult length at birth, and 50 per cent of adult height at two years, the famous measuring point most parents know about (see pls xvi, xvii).

The *Massospondylus* embryo highlights some interesting facts about growth. Just as with humans, the head is relatively large, the eyes are

The embryo of *Massospondylus* curled up neatly inside its egg.

also large (this makes babies look cute to their parents), and the neck and tail are unusually short. As *Massospondylus* grew, its head and eyes grew larger more slowly than the rest of the body. On the other hand, the neck and tail sprouted in length much faster than the growth in length of the torso. Finally, the arms and legs were already strongly built in the embryo – it looks as if these little dinosaurs might have been ready to run the minute they were born, just like a baby deer or a calf today. Such offspring are up and ready to run within minutes of birth, whereas human babies have short and weak arms and legs, and they cannot support their weight for many months.

Genus:	**_Massospondylus_**
Species:	_carinatus_

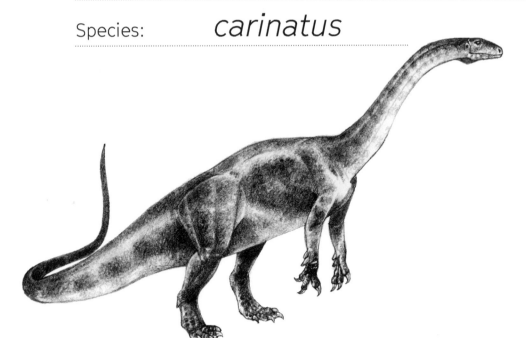

The solution to the dilemma of delicate bones and needle damage was to scan the embryos. The team took one of the _Massospondylus_ embryos to the European Synchrotron Radiation Facility (ESRF) for CT scanning. The buildings of the ESRF are on the banks of the Rivers Drac and Isère, on the edge of the city of Grenoble, dominated by a ring-shaped structure, 844 metres (2,769 feet) across, and housing a linear accelerator gun that produces the most powerful X-rays in the world. The beam is tapped off through forty-four beamlines and used for thousands of experiments in all branches of science each year. The team of palaeontologists made ultra-high-resolution scans of the tiny embryo.

The scan (see pl. xvi) shows a slightly flattened skull, but with all the component bones present, and highlighted in different, bright colours after processing. The scan confirms that this embryo had a full set of teeth, with rather long, sharp incisors at the front, and broader cheek teeth behind. This fully developed set of teeth suggests that the embryo was ready to feed the minute it hatched out – no evidence of parental care here, but a well-developed baby that hatched out and staggered towards the nearest plant food it could find on day one. This confirms the evidence from the stocky little limbs – this dinosaur could look after itself as soon as it hatched.

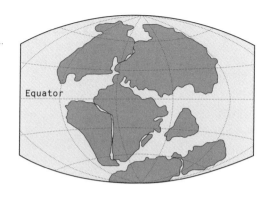

Named by:	Richard Owen, 1854
Age:	Early Jurassic, 201–191 million years ago
Fossil location:	South Africa, Zimbabwe
Classification:	Dinosauria: Saurischia: Sauropodomorpha: Massospondylidae
Length:	4 m (13 ft)
Weight:	490 kg (1,080 lbs)
Little-known fact:	A close relative, *Sarahsaurus*, is known from Arizona in the United States.

In 2018, two scientists from the Bernard Price Institute in Johannesburg published a full description of the CT-scanned skull[1] of an adult *Massospondylus* (see pl. xvii), and this time they did not have to fly to Grenoble as the University of Witwatersrand had purchased its own CT scanner. The skull shows how the eye socket, though huge, is relatively smaller than in the baby, but the snout is about the same length (babies often have short snouts). We now have all the detail of baby and mother.

What about dinosaur nests and parental care?

Did dinosaurs care for their young? The fact that they laid lots of eggs and that their young could fend for themselves as soon as they hatched suggests they did not look after their offspring. The mother cod does not do much for her young, simply dumping the eggs and leaving, but other

---

1 The scan files have been made available, so anyone can laser-print a 3D *Massospondylus* skull if they wish: https://3dprint.com/200131/dinosaur-fossil-3d-scanning/.

fish do protect their eggs. Some brood the young in their mouths, and the male seahorse famously has a brood pouch in which up to 2,000 miniature seahorses hatch and grow up. Crocodiles and turtles lay their eggs in safe spots on the beach or on the banks of rivers. We've all seen film of the mother sea turtle hauling herself laboriously up the beach, digging out a deep pit, laying her eggs, and covering them. Then she (and of course the father) are nowhere to be seen when the young hatch and scuttle down the beach.

Crocodiles and alligators construct a bowl-like nest in the mud on the shore of their river, lay between ten and forty-five eggs, and cover them with soil and leaves. They generally don't disappear, however, and naturalists were amazed in the 1960s to confirm that crocodiles and alligators offer some parental care, belying their Victorian image as brutal creatures that were best observed in the form of handbags. The mother and father defend their riverbank territory from predators, including (in Florida) raccoons, which will gladly eat all the eggs in a nest. As they come near hatching, the young alligators chirrup to each other and to their parents.

Genus: **Maiasaura**

Species: *carinatus*

The young bash their way out of the egg using a special egg-tooth on the snout (as do birds, and so almost certainly dinosaurs had the egg-tooth too) and the parents will stick around to defend their young from predators. The mother often carries her hatchlings down to the water in her mouth, where they quickly learn to catch small prey such as snails, insects, tadpoles, minnows, and crayfish. She will protect them for two years, at which point she chases them off and prepares to produce a new egg clutch. When early naturalists saw crocodiles and alligators carrying their young to the water, they declared (of course) that these beastly reptiles were eating their own young.

Determining whether dinosaurs cared for their young or not is almost a political thing. The Victorian view would have been 'not'. Dinosaurs were brute beasts to them, like the evil modern crocodiles, and so it was taken for granted that the ancient animals had no parental instincts at all. Then entered a revisionist view in the 1970s, linked with new views about dinosaurs as warm-blooded and active, and dinosaurs suddenly became loving and gentle. The heroine was *Maiasaura*, meaning 'good mother reptile', named from abundant remains in the Late Cretaceous of

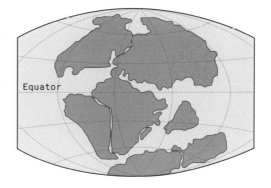

Named by:	**Jack Horner and Robert Makela, 1979**
Age:	**Late Cretaceous, 80–75 million years ago**
Fossil location:	**Montana, USA**
Classification:	**Dinosauria: Ornithischia: Hadrosauridae**
Length:	**9 m (30 ft)**
Weight:	**4–5 tonnes (8,800–11,000 lbs)**
Little-known fact:	**Most focus is on the female *Maiasaura*, but males may have head-butted when seeking mates, using spiky crests in front of their eyes.**

Montana that had been found in association with numerous nests and eggs. At the time, it was argued that whole gangs of *Maiasaura* mothers congregated to lay their eggs in nest mounds, and they sat there, chirping to each other, and spaced out enough to be sociable, but not so close that they would interfere with each other. The nesting sites showed stacks of nests, accumulated over many years, so the dinosaurs seemed to show nest-site fidelity, coming back year after year. Most importantly, these dinosaurian Earth mothers were said to have cared for their young, collecting succulent planty morsels for them, and hanging around the nest site as the babies took their first steps (see pl. xiv).

The original authors have stepped back from some of the more extreme interpretations. Levels of proof must be maintained, however attractive the idea of loving and laid-back dinosaurs, and some of the assumptions were hard to demonstrate with fossil evidence. Much more convincing is evidence from those so-called *Protoceratops* nests from Mongolia first reported by Roy Chapman Andrews in the 1920s. It turns out that the palaeontologists back in the 1920s had got it all wrong.

In 1993, in a second series of AMNH expeditions to Mongolia, Mark Norell and colleagues found further nests like those reported in the 1920s, but this time they examined the eggs in more detail. The palaeontologists found to their surprise that tiny bones inside some of the eggs belonged to a flesh-eating theropod dinosaur, not *Protoceratops* at all. It turns out that the 'egg thief' **Oviraptor** had been sorely misjudged; it was hanging around the nests because they were its own nests.

Even more remarkable, Norell found a complete skeleton of an *Oviraptor* parent, apparently incubating the eggs. The presumed mother had her legs tucked underneath her body, running between two half circles of eggs, and her arms stretched round and back at her sides. In life, she was covered in feathers, and so she was clearly incubating her eggs in the ground nest, just as an ostrich does today. Presumably, she stepped into the middle of the nest, being careful not to crush any of her eggs, and then folded her legs while shuffling eggs out of the way, and then settled down, and flopped her feathered arms out over the egg rows on either side.

This is a great story in which further research has corrected an earlier misconception – not so much about whose eggs they really were, but more that the parents incubated their eggs. This is very bird-like behaviour, and it could not have been assumed in dinosaurs – they might very well have laid their eggs and covered them with soil and a compost of leaves, as crocodiles do, and then more or less abandoned them.

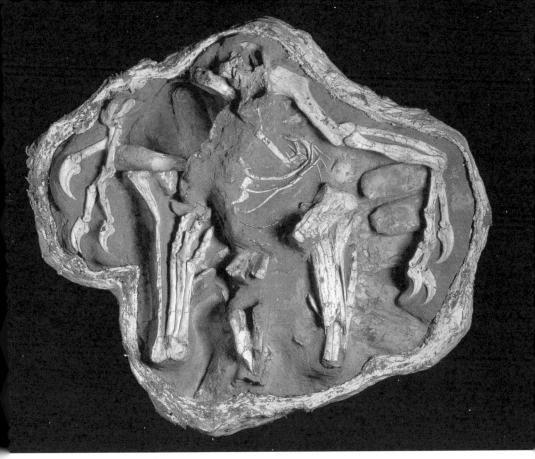

The famous skeleton of a mother *Oviraptor* incubating her eggs.

Now we know that at least the small and medium-sized theropod dinosaurs, those that happen to be most closely related to birds, did incubate their eggs in a bird-like way.

Did they engage in further parental care? The well-developed teeth of *Massospondylus* suggested that the babies were self-sufficient on hatching, so speaking against parental care in those dinosaurs; and such well-developed, even slightly worn, teeth have been noted in other dinosaur embryos. But then, sometimes whole clusters of dinosaurs are found together, most famously in the case of a horn-faced plant-eater, the ceratopsian **Psittacosaurus** (see overleaf and pl. xv). Hundreds of clusters of juvenile *Psittacosaurus* have been found in the Early Cretaceous rocks of north China – but this may simply represent groups of young hanging out together for safety, and it's not clear whether their parents were there to supervise the crèche, except in the case of one disputed specimen where an adult *Psittacosaurus* skull appears to have been attached to a cluster of twenty or so babies.

Genus: **Oviraptor**

Species: *philoceratops*

Genus: **Psittacosaurus**

Species: *mongoliensis*

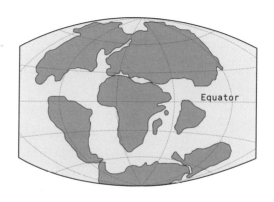

Named by:	**Henry Osborn, 1924**
Age:	**Late Cretaceous, 76–72 million years ago**
Fossil location:	**Mongolia**
Classification:	**Dinosauria: Saurischia: Theropoda:**
	Oviraptoridae
Length:	**2 m (6½ ft)**
Weight:	**20–30 kg (44–66 lbs)**
Little-known fact:	***Oviraptor* had a crest on its short snout,**
	possibly coloured in life and used in display.

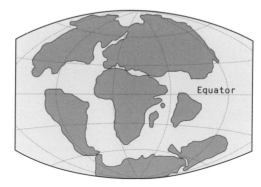

Named by:	**Henry Osborn, 1923**
Age:	**Early Cretaceous, 125–100 million years ago**
Fossil location:	**Mongolia, China**
Classification:	**Dinosauria: Ornithischia: Ceratopsia:**
	Psittacosauridae
Length:	**2 m (6½ ft)**
Weight:	**40 kg (88 lbs)**
Little-known fact:	**This is the most abundant of all dinosaurs,**
	with thousands of specimens already found
	in northern China.

How fast did dinosaurs grow up?

If you start small and end up huge you must either grow very fast or live for a very long time. This was the conundrum that Greg Erickson decided to tackle early in his career. He has shown that dinosaurs generally grew to adult size really fast, and this is another aspect of how they could be so huge.

The evidence comes from growth rates in the bone. As we saw in Chapter 4, dinosaurian bone represents a midway stage between the bone of most modern reptiles and most modern mammals. In thin section, dinosaur bone can show all the details seen in a thin section of modern bone, so it's possible to make direct comparisons and without any fear that details have been crushed or remodelled. In fact, dinosaur bone sections can be very mammal-like in showing high-energy open-weave bone structures with evidence of secondary remodelling as minerals were recycled. But, at the same time, in certain bones of the skeleton, very clear growth rings are seen. These, as in modern reptiles, and indeed in trees, are light and spacious when growth is rapid (usually summer), and tight and dark when growth is slow (usually winter), when life conditions are poor.

In a series of papers, Greg Erickson has explored the growth rates of many dinosaurs using observations of their growth rings. In a classic study of *T. rex* and its relatives, Erickson counted growth rings in bones from animals of all sizes. In one example, he counted to nineteen growth rings, and he was confident he had reached the endpoint because the outside of the bone was finished with some tightly packed bone layers, called the external fundamental system (EFS). (The true age at which the animal had died was twenty-eight years, when growth rings effaced by remodelling were accounted for.) Getting access to the bones took a long time. 'Curators initially weren't too enthusiastic about my requests to cut their treasured dinosaurs into pieces,' he notes. 'But things have loosened up a bit. Successes in ageing specimens by my research team and by others have made access much easier today than it was in the past.' Over years, he accumulated enough examples to be able to plot the growth curves for *T. rex* and its close relatives.

Erickson found that all the tyrannosaur growth curves were S-shaped, with slow rates of growth for the first five years, then rapid increase in size till the ages of fourteen to eighteen, and a levelling off at that point. The final point of adulthood varies by species, from age thirteen to fifteen in the smaller tyrannosaurs *Albertosaurus*, *Gorgosaurus*, and *Daspletosaurus*

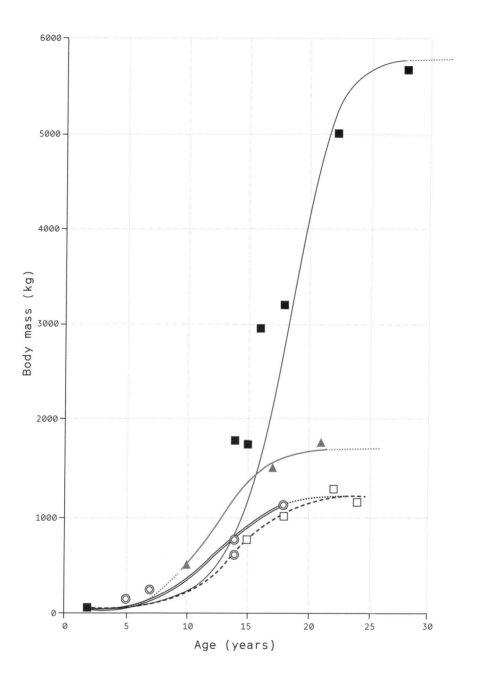

Graph showing the estimated growth curves of different specimens.

Maximum growth rates:

— ■ *Tyrannosaurus* = 767 kg y⁻¹
— ▲ *Daspletosaurus* = 180 kg y⁻¹
═ ◎ *Gorgosaurus* = 114 kg y⁻¹
--- □ *Albertosaurus* = 122 kg y⁻¹

to twenty to twenty-five years in *Tyrannosaurus* itself – this is when growth slows down. This suggests that *T. rex*, for example, grew from a 1- to 2-kilogram baby to a 6-tonne adult very fast, and it put on most weight between ages fourteen and eighteen, so it added about half a tonne per year during that period of sustained growth.

This is a contentious subject, and some critics argue that the growth rings are not always annual. For example, if food supplies are abundant and the winter is mild, perhaps, they argue, you would not get the dense, winter ring when growth normally slows. Or, on the other hand, a time of freak weather conditions when plants disappeared or the weather became stormy in summer might add an extra dense, slow-growth ring. Greg Erickson responds that:

> ...all of this might be true, but modern reptiles show pretty good matching of growth rings and age regardless of environmental conditions. Furthermore, reptiles raised under constant conditions or from non-seasonal environments still lay down single annual lines. This suggests the rings are primarily the result of annual developmental fluctuations in growth, rather than being caused by sporadic climatic fluctuations. We probably get it right for dinosaurs, most of the time.

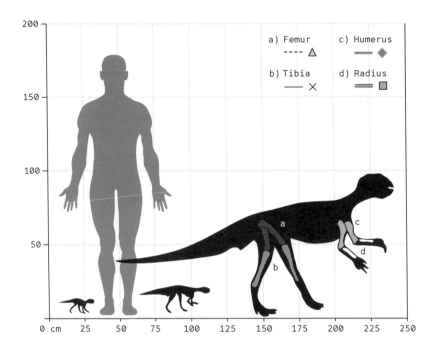

He also notes that 'in some cases we were able to cross-check multiple specimens of dinosaurs of the same size, and their ages estimated from the numbers of growth rings were always in the same ball park'.

Other researchers have applied the growth-ring method. For example, Qi Zhao, then my doctoral student, explored bone growth in his favourite dinosaur, *Psittacosaurus*. This is a plant-eating dinosaur, about 2 metres (6½ feet) long, that is very commonly found in the Early Cretaceous

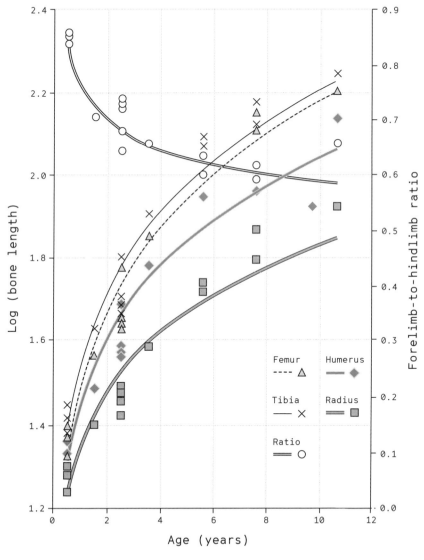

Posture shift (opposite) and growth
curve (above) in *Psittacosaurus*.

rocks of northern China, and many other parts of Asia. It has a short, blocky head, with a blunt nose and well-developed beak for chopping tough plants, and was bipedal as an adult, with strong arms and legs. The juveniles, on the other hand, were tiny, maybe 10–20 centimetres (4–8 inches) long, and quadrupedal. Qi used bone rings to work out the ages of each little baby in one of the juvenile crèche specimens (see pl. xv). Much to his surprise, he found that five of the six little chaps were only two years old, whereas a sixth was three years old – we usually assumed that such clusters of young would all have been the same age. So, these little dinosaurs, perhaps siblings, clustered together for protection from predators, and then, in this case, were swamped by ash fall, in the famous Chinese locality of Lujiatun in Liaoning province, sometimes called the 'Chinese Pompeii'.

Another interesting discovery about *Psittacosaurus* was that it underwent a posture shift at about the age of three, from being quadrupedal as an infant to being bipedal as an adult. On measuring the lengths of key limb bones from a series of specimens, Qi matched these with their estimated ages. This analysis showed that the animal grew fast in its first four years, doubling in length each year, and then reached adult size by about the age of seven or eight. This modest-sized dinosaur was ready to breed by the age of six or seven. Like all dinosaurs, and most modern reptiles, it continued to grow more or less forever – this accounts for travellers' tales of rare, absolutely huge, ancient crocodiles or snakes. Most animals die younger. By contrast, mammals and birds cease growing in height or length soon after sexual maturity.

In his studies of different dinosaurs, Martin Sander of the University of Bonn found the greatest age in the sauropod *Janenschia*, which took about fifty years to reach its full 20-tonne (44,100-pound) size. Despite earlier guesses, no bone histological evidence suggests any dinosaur reached an age of 100 years, and they were all breeding much earlier than that. This makes biological sense in terms of evolution and the need to breed as soon as reasonably possible.

So far, we have established that dinosaurs shared four unique features – bird-like respiration and efficient use of oxygen, warm-bloodedness because of their large size, a mixed quantity-quality reproductive mode (laying lots of eggs; small eggs; scant parental care), and relatively fast growth to adult size. These four characteristics seem to be smart ways for dinosaurs to save energy, but can they work together to explain in some way how dinosaurs could have achieved the truly colossal sizes that they did?

How could dinosaurs be so huge?

I am often asked 'what is the point of palaeontology?' I usually mumble something about origins and evolution, and wider cultural understanding of the history of life and Earth's environments. One key reason, though, is that some ancient organisms broke all the rules. Biologists say that elephants are about as big as a land animal can be without collapsing under the weight of too much flesh, or starving to death when the climate changes. So, the dinosaurs, and sauropods in particular, are a fine example of the impossible being real – we can't say that gravity was lower in the Jurassic, or that they spent their entire lives under water (even though some crackpots make these claims). So, these huge sauropods really existed and they were really too big to be true. How on Earth can we explain that?

The question has been solved thanks to the combined efforts of many palaeobiologists, focused by Martin Sander. He had a big idea, and he raised 5 million euros for a long-running research project from 2004–2015, and one that every school child might dream of – his project was titled 'Biology of the sauropod dinosaurs: the evolution of gigantism'. Sander recruited twenty or more researchers, not just palaeontologists, but also experts on nutrition, botanists, and zoo keepers. He wanted to resolve once and for all just why the sauropods were so huge.

He had in mind the largest dinosaur of all time, *Brachiosaurus* (see overleaf) from the Late Jurassic of Tanzania, in east Africa, and the midwestern United States. Its skeleton was a staggering 26 metres (85 feet) long, equivalent to two regular coaches parked nose to tail, and it raised its head to as much as 9 metres (30 feet) above the ground, the height of a three-storey building. Unlike other sauropods, *Brachiosaurus* had extra-long front legs, which raised the front half of its body, somewhat like a giraffe, and the vertebrae of the neck show that the natural position of the neck was at about a 45-degree slope. In other sauropods such as *Diplodocus* and *Camarasaurus*, the neck was held more horizontal. So, the point of Sander's study was to work out how these 40- to 50-tonne monsters could function.

I went to one of the consortium meetings in Bonn in 2011 and was fascinated to hear about experiments on human physiology in which American professors had recruited students to survive on a range of bizarre diets, such as nothing but burgers or cabbage for a month (experiments that probably would not be permitted today), and the zoo keepers who measure the inputs and outputs of elephants and other

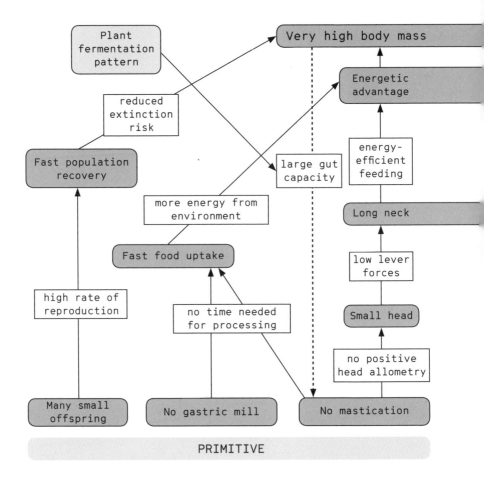

PRIMITIVE

beasts in their charge. The zoo keepers reported that an elephant must eat up to 270 kilograms (600 pounds) of forage each day. As Sander noted, if sauropods had the same physiology as a modern elephant, they would require ten times as much, in other words 2.7 tonnes (6,000 pounds). That's a pile of leaves as big as a passenger coach, each day. Further, noted the zoo keepers, rolling their eyes to the heavens, their elephants turn those 270 kilograms of plant food into 70 kilograms (over 150 pounds) of dung each day – that's several dozen wheelbarrows-full.

Sander wanted to know what food plants were available to sauropods in the Mesozoic, and how sauropod physiology differed from that of elephants. Of course, their bone histology indicates warm-bloodedness, but they were so huge, at 50 tonnes or so, that if they fed at the rate of an elephant, they could not have packed away enough plant food through their tiny heads and long necks. So, he wove together what we know about dinosaurs in general, and sauropods in particular, to develop an

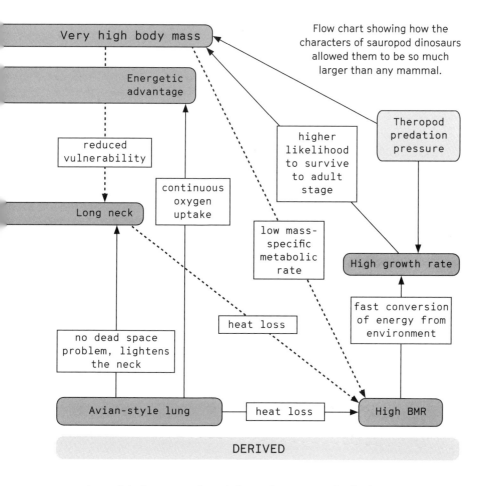

Flow chart showing how the characters of sauropod dinosaurs allowed them to be so much larger than any mammal.

Very high body mass

Energetic advantage

reduced vulnerability

higher likelihood to survive to adult stage

Theropod predation pressure

continuous oxygen uptake

Long neck

low mass-specific metabolic rate

High growth rate

no dead space problem, lightens the neck

heat loss

fast conversion of energy from environment

Avian-style lung

heat loss

High BMR

DERIVED

overview of their secret – here is how the sauropods, the largest animals of all time, apparently did the impossible.

It was a combination of many small offspring and small eggs but no parental care; small head and no chewing; and bird-like lungs, which processed oxygen pick-up more efficiently than reptile and mammal lungs. These characteristics allowed sauropods to achieve huge size for minimal food intake – probably as much as an elephant, or even less, for a body that was ten times as large. They achieved steady body temperature by being huge, not by eating lots and having complex inner furnaces, as elephants and humans do. They laid small eggs and abandoned them, unlike elephants and humans, who spend a huge amount of time and energy caring for one or two babies, which can exhaust the mother's reserves of food. Martin Sander's spider diagram explains it all in a very convincing way – this is how sauropods broke free from the constraints that limit elephant, and mammal, size.

Genus:	**_Brachiosaurus_**
Species:	_altithorax_

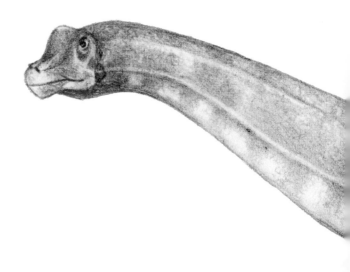

Named by:	**Elmer Riggs, 1903**
Age:	**Late Jurassic, 157–152 million years ago**
Fossil location:	**United States, Tanzania**
Classification:	**Dinosauria: Saurischia: Sauropodomorpha: Brachiosauridae**
Length:	**26 m (85 ft)**
Weight:	**58 tonnes (127,868 lbs)**
Little-known fact:	**The skeleton of _Brachiosaurus_ from Tanzania on show in Berlin's Humboldt Museum is the largest dinosaur on show anywhere, standing 9 m (30 ft) tall.**

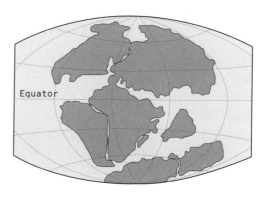

Equator

Were there ever dwarf dinosaurs?

Having achieved huge sizes, why would dinosaurs become small? One lineage, the maniraptoran theropods, became smaller and smaller, and sprouted long arms to improve their adaptations to tree-dwelling and eventually flight (see Chapters 4 and 8). Their new mode of life, hopping about in the trees, can explain the move to small size. Here and there, some dinosaurs became small because they lived on islands. The most famous are the dwarf dinosaurs of Transylvania – which sounds a bit like a movie title, but is not. These dwarf dinosaurs did indeed live in the corner of Romania classically called Transylvania, and they were reported first by Baron Franz Nopcsa, who was an impoverished nobleman from what was then (at the end of the nineteenth century) part of the Austro-Hungarian empire.

I first did research in Romania in 1993, four years after that country had overthrown the Soviet-oriented government in a violent uprising – I was shown bullet holes on some of the buildings of Bucharest University. The Discovery Channel were keen to film a programme about Nopcsa, mainly because of his colourful life – he was not only a nobleman, but was also gay, and travelled Europe with his faithful secretary and lover, Bajazid Doda. Nopcsa was multilingual and spoke about his dinosaur work at conferences in England, France, and Germany, and had to sell fossil collections to remain financially solvent. He spied for both sides during the First World War, the Austro-Hungarian empire and Britain, worked with the Albanian partisans, and offered himself to be King of Albania. Eventually, in poverty and despair, he shot Doda and himself in 1933. This was deemed to be about enough for a thirty-minute film, but I kept insisting we needed some science, and that the dwarf dinosaurs were really something biologically important.

Nopcsa was the first to note that his Transylvanian dinosaurs were dwarfed, at a meeting in Vienna in 1912. He observed that the Transylvanian dinosaurs rarely exceeded 4 metres (13 feet) in length and the largest one, a sauropod later named ***Magyarosaurus*** *dacus* (see overleaf), was a puny 6 metres (20 feet) long compared to 15–20 metres (49–66 feet) for its closest relatives elsewhere. During the discussion following his paper, Othenio Abel, a great Austrian palaeobiologist, agreed and said the phenomenon was just like the dwarfing of elephants, hippopotamus, and deer on Mediterranean islands during the Ice Ages.

Between them, Nopcsa and Abel had nailed it. There have been many evolutionary explanations for the phenomenon, but it is surely mostly

Baron Franz Nopcsa in the costume
of an Albanian freedom-fighter.

to do with the fact that islands support fewer species and have simpler ecosystems than comparable sections of the mainland. So, with fewer species, less food, and smaller range sizes, animals can adapt their sizes, diets, and habits, and large forms have to become smaller. The pygmy elephants of the last million years of the Mediterranean, on islands such as Malta, Sicily, and Sardinia, were only 50 centimetres to 1 metre tall (1½–3 feet) at the shoulder, compared to 4–5 metres (13–16½ feet) for an adult elephant today. Evidently, elephants, hippos, and other African mammals got across to these islands at times when Mediterranean sea levels were much lower than they are today because water was locked up in huge northern ice caps.

The Transylvanian dwarf dinosaurs lived on Haţeg island, which measured 100–200 kilometres (62–124 miles) across, and was one of several large islands in the Late Cretaceous, when sea levels were very high and flooded most of the southern parts of Europe. Studies of the bone histology of three of the dwarf dinosaurs, the sauropod *Magyarosaurus* and the ornithopods *Telmatosaurus* and *Zalmoxes*, show that these were adults, not juveniles. They were all from one-third to

Genus:	**_Magyarosaurus_**
Species:	_dacus_

Named by:	**Friedrich von Huene, 1932**
Age:	**Late Cretaceous, 72–66 million years ago**
Fossil location:	**Romania**
Classification:	**Dinosauria: Saurischia: Sauropodomorpha: Titanosauridae**
Length:	**6 m (20 ft)**
Weight:	**0.75 tonnes (1,654 lbs)**
Little-known fact:	**_Magyarosaurus_ is an 'island dwarf', much smaller than related sauropods, because it lived on the Haţeg island of Transylvania.**

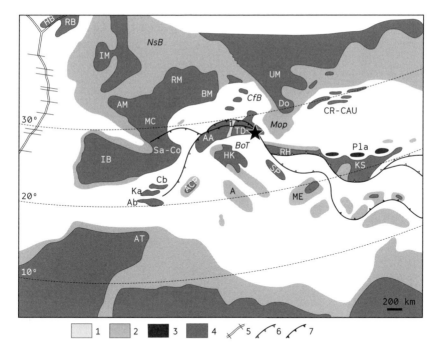

Palaeogeography of Europe in the Late Cretaceous, showing how southern and eastern Europe were divided into many islands because of high sea levels at the time (Haţeg island is marked by a black star).

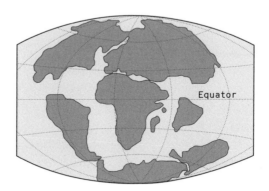

one-half of the body length of their nearest relatives from mainland areas in Europe and North America.

Not only had they shrunk in size, but they seemed somehow to be 'primitive', dinosaurs whose closest relatives were known from mainland spots 20–30 million years older. Presumably their ancestors had become established on the coastal area and then cut off as sea levels rose. Then, while their nearest relatives on the mainland continued to evolve and change, the island forms carried on as they were, in less complex ecosystems, and perhaps not subject to the same competitive pressures.

So, mostly dinosaurs grew larger and larger through evolutionary time, except in the case of these rare island forms. It's interesting to know that dinosaurs showed just the same adaptive capabilities as mammals, and they could become small when it was advantageous in evolutionary terms.

How the dinosaurs achieved giant size is one of the key conundrums in palaeontology. In fact, it has become almost a philosophical question to consider what was special about dinosaurs in general, and the giant sauropods in particular, that enabled them to be ten times as big as an elephant. The philosophical issue is that the dinosaurs stretch our understanding of what is possible in physiology and in evolution. The question also reminds us to keep a firm control on speculation and not to suggest, for example, that the Jurassic monsters lived underwater or that gravity was lower in the past.

Determining growth rates from close study of dinosaurian bone structure and growth rings has been a major advance. In a 2017 paper, Greg Erickson and colleagues applied growth-ring analysis to the teeth of embryo dinosaurs, and they found evidence that development in the egg was slow, lasting perhaps two to six months, at rates akin to modern reptiles rather than the much faster rates of modern birds, whose embryos develop from fertilization to hatching in eleven days to three months. Each of these steps in research knowledge is slow, and the work often painstaking. The trickier the work, the harsher the criticisms, but that is a manifestation of the self-correcting property of science.

What's still to be discovered in this field? Greg Erickson reflects on what might come next:

The new generation of palaeobiologists have made great headway in establishing methods to study the life history of dinosaurs and bring these animals back to life. Reconstructing age-mass growth curves was pivotal. This allowed quantification for how these animals developed, leading the field from speculation into the realm of science. Growth curves enabled standardized direct comparisons of dinosaur growth to those of living animals, and with one another. This provided inroads to understanding myriad aspects of dinosaurian biology including links between growth and evolution, physiology, reproduction, population biology, and even the evolution of the modern bird characters. You name it! There are many more species to be studied and we need more cross-checks from those that have been studied to add statistical power. The advent of modern high-resolution imaging techniques, including synchrotron X-ray imaging, hold great promise for allowing rapid, non-destructive analyses of dinosaur growth. These will speed up our accumulation of knowledge and allow access to rare specimens that curators are still reluctant to provide for destructive sampling.

Chapter 7

How Did Dinosaurs Eat?

Employing the software used to design skyscrapers to determine how dinosaur jaws worked might seem far-fetched. Yet our understanding of dinosaur feeding has been revolutionized by the application of an engineering design tool developed in the 1940s. The pioneer was Emily Rayfield, now a professor at the University of Bristol. She is a no-nonsense daughter of a Yorkshire pig-farmer, and brings that practical background to bear on her work: 'Once I had to borrow a couple of pig skulls from my father for a research project testing our computer models of bone strength,' she recalls. Rayfield is at the centre of a growing team of successful students she has mentored over the years in their engineering-based studies of dinosaur function.

When she began her doctorate in 1997 at the University of Cambridge, Rayfield was set the task of determining the feeding mechanics of **Allosaurus** (see overleaf), a large predator from the Late Jurassic Morrison Formation of North America. *Allosaurus* was known from many skeletons and skulls, and it was the top predator of its day, feeding on two-legged plant-eaters, as well as the iconic *Stegosaurus*, with its tiny head held low, the great arched back lined with a double row of bony plates along the midline, and its long tail that ended with four vertical spines.

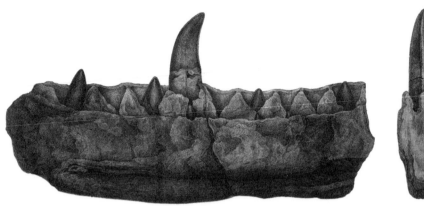

The lower jaw of *Megalosaurus*, the first dinosaur to be named, showing the sharp, knife-like teeth of a predator.

Allosaurus was a biped, some 8.5 metres (28 feet) long, and with massive hind legs for powerful trotting and short, but strong, arms with which it could manipulate its prey. It shared the role of top predator in the Morrison Formation with *Ceratosaurus*, a 5.7-metre-long (18¾-foot) biped that was remarkable for the heavy bony shelves and projections on top of its skull. *Allosaurus* had a high skull, with many openings for sensory organs and other structures, and strong struts between. Each

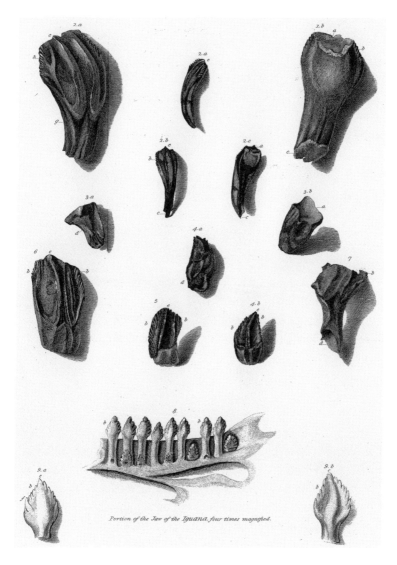

The teeth of *Iguanodon*, the second dinosaur
to be named, showing the blunt-edged, ridged
teeth of a herbivore.

Genus:	*Allosaurus*
Species:	*fragilis*

jaw was lined with between fourteen and seventeen scimitar-like teeth, each 6 centimetres (2½ inches) long, pointed and curved, and with jagged edges front and back. This is a classic design in predatory dinosaurs, with the teeth curved back to hold the prey animal firm and make sure that if it struggles it is forced backwards down the throat.

Emily Rayfield was motivated by the desire to develop new methods to shift the study of dinosaur feeding from speculation to testable science. As she says, 'The basics of how dinosaurs fed had been known for 200 years ever since the first discoveries of dinosaurs, and not much had changed since then.' For example, *Megalosaurus* from the Middle Jurassic

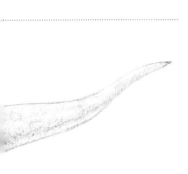
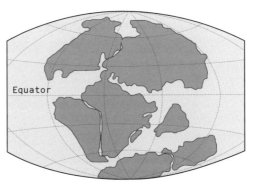

Equator

Named by:	**Othniel Marsh, 1877**
Age:	**Late Jurassic, 157–152 million years ago**
Fossil location:	**United States, Tanzania, Portugal**
Classification:	**Dinosauria: Saurischia: Theropoda: Allosauridae**
Length:	**8.5 m (28 ft)**
Weight:	**2.5 tonnes (5,513 lbs)**
Little-known fact:	**One *Allosaurus* individual from Wyoming had nineteen broken bones, some of which had begun to heal, but others were infected – he suffered for six months before he died.**

of England, named in 1824, showed the same sharp, scimitar-like teeth as in *Allosaurus*, and from the start it had been identified as a meat-eater, from a comparison with modern crocodiles. On the other hand, *Iguanodon* from the Early Cretaceous of England, the second dinosaur ever to be named, had huge, blunt-ended teeth, which were compared with those of the modern iguana, a plant-eating lizard. In coming to their conclusions about diet, the early dinosaur scientists were using the classic principle of 'analogy with living forms' to make their functional interpretations. In other words, they were applying a common-sense assumption that conditions in the past were uniform and concordant with those today.

Emily Rayfield, the great innovator in
engineering approaches to dinosaur
feeding function, in her laboratory.

We assume that teeth in the Jurassic and Cretaceous share functional
characteristics with those today, even if we are looking at animals, such as
dinosaurs, without obvious modern relatives that shared their lifestyles.

Emily Rayfield had this basic information, but she wanted to know
more. Her solution was to apply an engineering method called finite
element analysis (FEA) to her dinosaur question. The FEA method had
been developed in the 1940s as a tool for engineers and architects to
improve the efficiency of their structures. Instead of following medieval
practice, and making buildings super-massive so they wouldn't fall down,
FEA was a way for architects to stress-test accurate models digitally
before construction began. The trick is to make the 3D model and then to
apply the correct material properties, such as how bendy the material is
under tension and compression, how elastic it is, how it deforms under
opposing forces, and its density.

During the next three years, Rayfield had to sweat to apply the FEA
method to the skull of *Allosaurus*. 'If I had failed to find a way to make
the FEA software read in the details of the detailed structure of the
dinosaur skull, I wouldn't have had a thesis to present to my doctoral
committee,' she recalls. Nobody had done this before, and she was not
a trained software engineer. Nonetheless, after a great deal of effort,
forcing different software programs to talk to each other, it worked.

She had figured out how to make a reliable engineering method tell us how dinosaur jaws worked.

Even getting the CT scans was a challenge. The skull of *Allosaurus* was in the Museum of the Rockies in Montana, and this 1-tonne (2,205-pound) specimen had to be moved by truck to the Deaconess Hospital in Bozeman, 3.2 kilometres (2 miles) away, where it was scanned. Then, rendering the 3D digital model and undistorting it to make an anatomically accurate model took more than a year. The image had to be passed from software package to software package, with the risk at every step of computer meltdown, or at least a string of error messages. Hardest of all was to read the model into the standard FEA software to complete the functional analysis. The 3D digital model of the skull had to be divided into geometric shapes, converting the skull into a kind of wire mesh model composed of numerous pyramids, called elements. Then each element was given material properties measured from the bone of modern animals, often pigs or cows because of the similarities in the internal structure of dinosaur and large mammal bone. Rayfield recalls:

> This was a stressful time, but the method worked. After that, I concentrated on making it more efficient, and advances in computing mean it goes much faster, and it's all a lot less risky. We combine our engineering studies with close observation of the fossils to find other clues about dinosaur function and behaviour and confirm our fossil models with studies on living animal bone function.

The methods she applied have led to remarkable new discoveries and astonishing precision in results, as we shall see.

Digital modelling and dinosaur bite force

We know FEA works because bridges and skyscrapers normally do not collapse, and aeroplanes on the whole manage to survive storms without falling to bits. So, the method presumably also works for dinosaurs. This is perhaps the nearest we can get to really testing aspects of the palaeobiology of dinosaurs. We are no longer guessing about how the bones functioned, but actually putting them to the test, using a digital model that is as near to the real thing as we can construct, and establishing hypotheses about function that are open to criticism and testing.

So, what was the maximum bite force of *Allosaurus*? Bite force is measured in newtons, with one newton equivalent to the force of holding a pencil. Tapping your boiled egg would take a force of about 35 newtons. Human bite forces are between 200 and 700 newtons, a lion is capable of 4,000 newtons, and the strongest bite force of all living animals is by the Great White shark, which can clamp down with a force of 18,000 newtons – thirty-six times the force humans can exert. Newtons of force can be converted to an equivalent mass, at roughly 9,800 newtons per tonne, so the Great White shark bite force is the same as applying a weight of about 1.8 tonnes. Rayfield's study showed that *Allosaurus* had a bite force of 35,000 newtons, much greater than any living predator.

Tyrannosaurus rex was even larger than *Allosaurus*, so it may have had an even more deadly bite. In a smart piece of independent confirmation, palaeontologists have homed in on this question from two directions. The first was through mechanical force tests in the laboratory. A piece of *Triceratops* bone was found with a deep gash on the surface. The researchers made a cast of the tooth puncture, and found it was identical to the tip of a *T. rex* tooth that had been driven in for 3 centimetres (1¼ inches) of its overall length of four times that. (Note that 12 centimetres is the length of the crown, the exposed part of the tooth – the whole tooth, including its root, was the length and shape of a regular banana, but distinctly sharpened at the business end.) The researchers then made a model *T. rex* tooth, mounted it on a pressure indenting rig, and drove it into pieces of cow bone. To match the 3-centimetre tooth pit took a force of 13,400 newtons, equivalent to 1.4 tonnes of weight applied. Was this the maximum possible bite force that *T. rex* was capable of?

Emily Rayfield tested the range of forces *T. rex* could have exerted within the limits of the strength of its skull and jaws, and came up with a maximum value of 31,000 newtons for the sum of values for each tooth, similar to the bite force of *Allosaurus*. In a further study, using a different biomechanical computing approach called multi-body dynamic modelling, Karl Bates and Peter Falkingham estimated a range of bite force values from 35,000 to 57,000 newtons, or equivalent to the application of 3.6 to 5.8 tonnes of weight. This is the strongest bite force ever demonstrated in any animal living or extinct, far higher than that of the living Great White shark, and far higher than the value calculated by the puncture experiment with the *Triceratops*; but then the *T. rex* who made that bite clearly wasn't really trying very hard. Importantly, all these different approaches give similar values, which suggests there's a good chance they might be correct. Further, it allows us to answer the classic dino-geek question: could *T. rex* have bitten a car in half? The answer is a resounding 'yes'.

In her comparison of three meat-eating theropods – the modestly sized *Coelophysis* from the Late Triassic, *Allosaurus* from the Late Jurassic, and *Tyrannosaurus* from the Late Cretaceous – Emily Rayfield showed they all employed different feeding methods (see pl. x). In *Tyrannosaurus*, peak stresses are in the snout, whereas in the other two, peak stresses are located further back, above the eye socket. This shows that *T. rex* used a puncture-pull means of killing and feeding, snapping with the front of its jaws to kill its prey, and then pulling back to tear the flesh from the carcass held down by its great foot. The other two theropods were capable of a more powerful bite along a greater stretch of the jaws, and so were perhaps juggling the prey in their mouths and chomping it into bits before swallowing.

In a later study of the strange, long-snouted theropod *Spinosaurus* from the Early Cretaceous of North Africa, Rayfield and colleagues

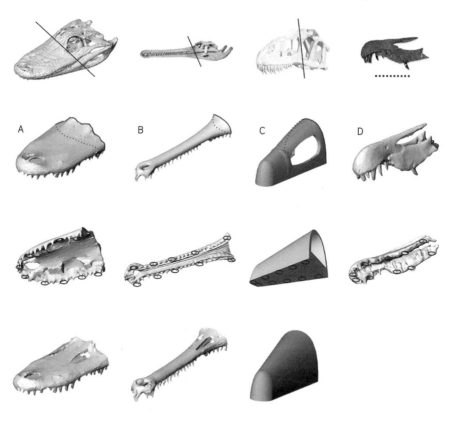

Establishing the function of the snout of the spinosaurid *Baryonyx* (D) with crocodile (A), gharial (B) and *Allosaurus* (C).

found that it functioned more like the skull of the modern gharial, or gavial (a slender-snouted fish-eating crocodile that patrols the River Ganges in northern India), than a crocodile or alligator, with their broad snouts. The researchers constructed 3D models of the snouts of all these animals, and then applied FEA. Crocodiles and alligators today employ twist-feeding, in which they sink their jaws into their prey, say a wildebeest snatched from the side of the river, drag it under the water, and then throw their body into great contortions. The rotation of the crocodile's body is transmitted to the side of its hapless prey, and a great chunk of flesh is twisted off. Gharials feed on fish by snapping their long delicate jaws shut and catching them in a cage of slender teeth. This is what spinosaurs did, and their bony palates acted to prevent their long snout from bending up and down, whereas in crocodiles and alligators the palate acts to strengthen the snout against twisting from side to side.

Fossil evidence for dinosaur diets

Tooth shape tells us whether a dinosaur was a herbivore or a carnivore. There are other kinds of fossil clues to diet, however. For example, palaeontologists look with sharp eyes for bite marks on bones, and dozens of examples have been published. Some of the biters can be identified rather precisely. If a *T. rex* bit into the bones of another dinosaur, it sometimes left scratch marks in parallel rows, and their spacing can be measured to test whether it matches the typical tooth spacing along a *T. rex* jaw.

There are also reports of stomach contents and even fossil excrement. The evidence of stomach contents can be contentious, as the discoverer has to convince others that the bones or stones or twigs inside the dinosaur rib cage really came from its stomach, and weren't simply washed in later by sedimentary processes after the animal died. On reconsideration of the evidence, it turns out that it was rare for the giant sauropods to swallow such stones, but small theropods related to birds frequently did. Modern birds, especially those that feed on tough plants, often swallow masses of small stones (gastroliths), which then reside in the crop, an expansion of the gastric system located above the stomach, and essentially take over the role of teeth in reducing the food to digestible bits. Mammals, of course, chew their food, and so few need gastroliths.

The metre-long coprolite of *T. rex*.

And then we come to coprolites, fossil faeces. The theologian and palaeontologist William Buckland, being English and privately educated, was fascinated by poop of all sorts. He was the first to describe coprolites formally, and he pictured a Jurassic ocean full of all the wonderful new marine reptiles that had been found by the famed fossil collector Mary Anning around 1820 on the Dorset coast – with each one shedding droppings into the ocean. Now, there are hundreds of reports of dinosaur coprolites, most of them probably accurate, but it can be hard to identify the culprit. One spectacular specimen – the granddaddy of all coprolites – was reported by Karen Chin in 1998. It's a 44-centimetre-long (17-inch) behemoth containing numerous bones of unidentified dinosaurs. As Chin said at the time, 'We're pretty sure this was dropped by *Tyrannosaurus*, but it can be tough to identify the poopetrator.'

Looking at teeth and coprolites is one thing, but such specimen-based studies do not help us to understand how dinosaurs actually fed. There are many more questions than simply, did they eat plants or flesh? For example, it would be good to understand whether particular herbivores fed on high or low plants, and whether they snipped or tore the food off, crushed or shredded it, and how much they ate every day. For carnivores, it would be good to know if they hunted actively or scavenged from dead carcasses, and whether they snapped at the flesh or sank their jaws in and twisted, as crocodiles do. Emily Rayfield's FEA work can resolve some of these questions, but the fundamental construction of the teeth themselves can be important.

Tooth engineering and plant-eating

So far, we have assumed that teeth are just teeth, but in fact they can be highly complex and exquisitely engineered tools. In our teeth, we have the classic three-part construction, a thin crystalline layer of protective enamel on the outside, and dentine, which forms most of the tooth and contains fine canals that connect to the nerves and blood vessels inside the pulp cavity. When the enamel is dissolved by bad food, as most of us can attest, our teeth become very sensitive and must be capped or repaired. If too much of the enamel and dentine is damaged, the whole tooth has to be removed, and this is not a trivial operation in humans because our teeth sit in deep sockets and they are held tight by cement.

Also, we generally want to hang onto our teeth because we only ever have two sets, our baby teeth and our adult teeth. In mammals, including humans, the teeth are replaced just once, when we are young, and we never grow any more. In fish and reptiles, on the other hand, the teeth are replaced dozens of times throughout their lives. If mammals had not reduced their tooth replacement to once only, we could eat what we liked, and there would be no need for brushing or flossing – or dentists, for that matter.

Dinosaur teeth can be enormously common as fossils. I remember vividly, as do many other palaeontologists, walking along the edges of the Sahara Desert, and picking up the large teeth of two predatory dinosaurs, *Spinosaurus* and *Carcharodontosaurus*, each about the length of the palm of your hand, and wondering at the sheer numbers of teeth those animals shed. This is like modern sharks, which are casting their worn teeth all the time, and predatory dinosaurs probably shed hundreds of teeth, getting through twenty or thirty full replacements in their lifetimes. In the dinosaur jaw, teeth line up below the socket, and new ones are always pushing the old ones out, sometimes even before they are worn away. This, though, is a part of the risk for an active predator, like a theropod or a shark – when *Allosaurus* was twisting and turning to subdue its prey, teeth got bent and torn from the mouth. Because new teeth grow into place sporadically, the jawline of a shark or dinosaur may be quite snaggly, with half the teeth more or less erupted from the gums, and others just poking through randomly along the jaws.

The most successful dinosaurs in terms of sheer numbers of individuals were the plant-eating hadrosaurs, commonly called duck-billed dinosaurs, because their long, horse-like skulls expanded at the front into a broad toothless structure, like a duck's bill. Hadrosaurs have been called the 'sheep of the Cretaceous', and in places, especially in North America and Mongolia, collectors often find hundreds of specimens together. Hadrosaurs had a remarkably standard skeleton and skull, but showed great diversity in their extraordinary head crests, different in each species. It's their teeth, however, that seem to have made them so successful. The huge success of hadrosaurs is hard to understand because stomach contents and coprolites show they were mainly eating conifers, and the tough twigs and needles of conifers seem pretty uncompromising as a food choice.

In a detailed study in 2012, dinosaur palaeobiologist Greg Erickson showed that hadrosaur teeth comprised six different tissues that worked together to make them remarkably effective and durable (see pl. xviii). Hadrosaurs were already famous for their huge numbers of teeth – as many as 2,000 altogether. Most of these were replacement teeth, lining up on the inside of the jawbone beneath the twenty to thirty functional teeth on each jaw margin. Their teeth were arranged in a straight line along each jaw, and tooth-to-tooth grinding kept them sharp. But it was the numerous different hard tissues that came as a surprise – just as in modern bison and elephants, the dental tissues were folded intricately so the harder enamel had maximum effect in tearing tough vegetation. Other tissues include forms of dentine and tooth cement, and giant filled tubules branching from the pulp cavity.

As the teeth ground each other down, while chopping tough conifer leaves, the six tooth tissues formed vertical structures that ground and chattered across each other, like a carpenter's rasp file. Erickson performed hardness and wear experiments on the fossil teeth and found the fossilized hard tissues responded as if they were fresh, providing values comparable with the most efficient grinding-toothed mammals today.

Hadrosaurs were successful even though – or maybe because – they tackled tough kinds of vegetation that, perhaps, other plant-eating dinosaurs could not manage. In effect they had bionic teeth, built like a steel rasp, and endless tooth replacement, so they could afford to let their teeth wear down quickly and then shed them. With such levels of wear, it would surely be useful if there were a smart way to divine the diet of dinosaurs by inspecting microscopic tooth wear patterns.

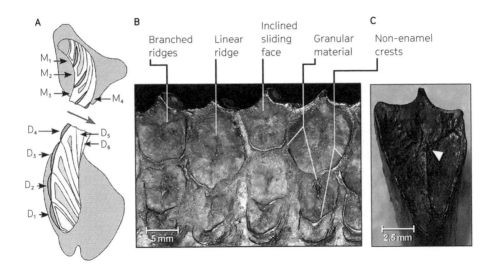

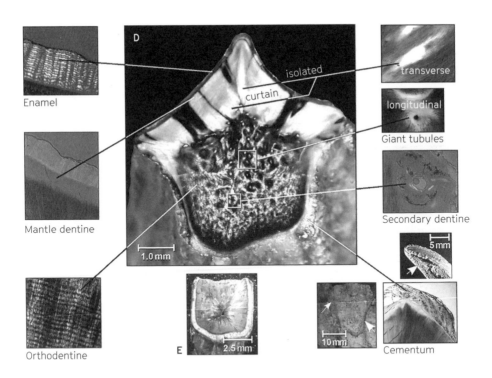

The unique teeth of hadrosaurs, showing how many teeth these dinosaurs had (B, C), and their multiple folds like those of a bison or horse.

Microwear on teeth can reveal diet

Palaeoanthropologists were the first to interpret microscopic pits and grooves on the teeth of early humans as clues to their diets, whether soft or tough plant materials, meat, or a mixture of both. In one of the textbooks from the 1970s, the instructor says that comparisons should be made between the fossil teeth and tooth wear in living animals. Indeed, experiments are described in which monkeys and apes are fed varied diets – maybe one eating cabbage, another eating grains, and another eating fruit – for a few weeks, and then their teeth are examined. In these enlightened times, we do not kill the animals, of course, but grip them between our knees, hold the mouth open, and insert some moulding compound to snatch an impression from their molar. In the textbook photographs, the monkey, held as in a vice between the anthropologist's knees, looks furious, and I wonder how exact the measurements could have been.

The real difficulty, however, is that every tooth surface is covered with pits and scratches, and identifying which ones were made by food items, and which ones by damage, can be a challenge. Especially with fossil teeth, it can be next-to-impossible to be confident that the marks were not made on teeth as they rolled along a river bed, or were otherwise beaten about. It is more reliable to take a micron-scale surface scan of the grinding surface of the tooth using a computerized microscope. The scan is then analysed automatically to classify all surface markings into sets, and the software is 'trained' so that, over time, it sorts out and discards random biffs and scrapes acquired during fossilization processes, and concentrates on saving information on plausible feeding marks.

Such studies of dinosaur tooth wear are in their infancy, but a really neat piece of work by Pam Gill, Emily Rayfield and their team discriminated diets of two of the earliest mammals, *Morganucodon* and *Kuehneotherium*, which scuttled and squeaked around the feet of the (probably oblivious) dinosaurs of their day. These two mammals are known from teeth and jaws, and some skeletal parts, from the Early Jurassic of South Wales, and their spiky little teeth suggested that they were insect-eaters. By mapping the key characters of modern bats, and their known diets, into a morphospace, the investigators determined that *Morganucodon* ate hard-cuticled insects such as beetles, whereas *Kuehneotherium* ate softer insects. This discovery was confirmed by FEA evidence of the jaw function, which showed that *Morganucodon* had a more powerful bite than *Kuehneotherium*.

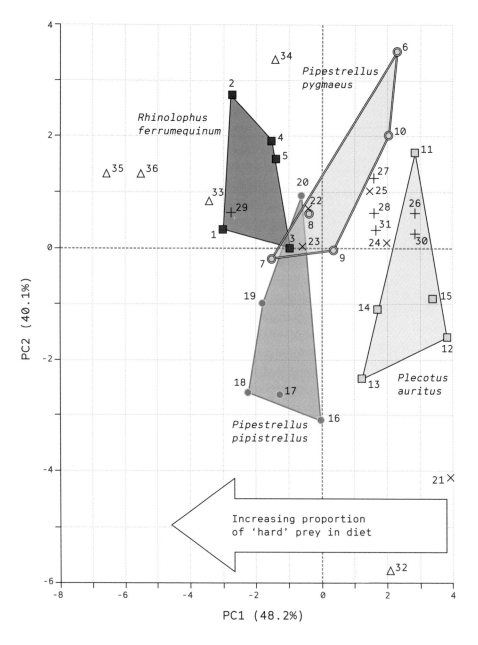

Morphospace showing tooth-wear characteristics among modern bats, and their different insect diets. By analysing fossils taxa such as *Morganucodon* and *Kuehneotherium*, their diets can be calibrated from the modern data.

△ *Morganucodon* (P3)

✕ *Kuehneotherium* (P3)

+ *Kuehneotherium* (P5)

There have been some studies on dinosaurian tooth wear, but they have been disputed. For example, the orientations of sets of scratches on the teeth in hadrosaurs were used to confirm the principal motions of jaw action, but older papers that sought to identify the precise diet from such scratches are now generally disavowed. A key challenge will be to identify appropriate modern analogues.

Dinosaur food webs

All species are interlocked in complex relations between predators and prey, and between species that compete for other resources. The food web, an example of which we saw for Wealden dinosaurs in Chapter 2, is a standard device used by ecologists to document these interactions, and to calculate the movement of energy from the sun as it is captured by green plants, which are then eaten by herbivores, which in turn are eaten by carnivores; and then all eventually die and rot away, releasing energy and carbon through detritus-eating insects and microbes. Food webs are usually shown as a sort of spider diagram that links predators and prey: fox eats rabbit eats grass. The top, or apex, predator (lion, killer whale, *T. rex*) is at the top of the diagram, and all arrows lead back to it. Dinosaur food webs can be very different from anything we see today.

In the example shown here, representing a particularly well-studied dinosaur fauna from the Late Cretaceous Adamantina Formation of Argentina, the top predators (number 1) are large theropods, including *Carnotaurus*, an abelisaurid with large head and short arms, carcharodontosaurids, and *Megaraptor*, a great name meaning 'big hunter'. In places there are three, four, or five steps from the bottom of the diagram up to these giant predators. Bottom of the pile are a bunch of fishes, frogs, a turtle, and beetles. The beetles are eaten by mammals, birds, lizards, snakes, and the weird armoured crocodile *Armadillosuchus*. The fishes are eaten by another crocodile, *Barreirosuchus*.

The strangest thing about this food web is the dominance by crocodiles (all shown in black in the diagram). Some were like modern crocodiles, presumably lurking in and around the rivers, and feeding on fishes and land animals they could snatch from the river banks. But others were much more adapted to life on land, with long legs, moving upright like a dog or hyena, and with short snouts and variable teeth, reflecting their different dietary adaptations. Some were active predators, even hunting dinosaurs, although probably only snatching at

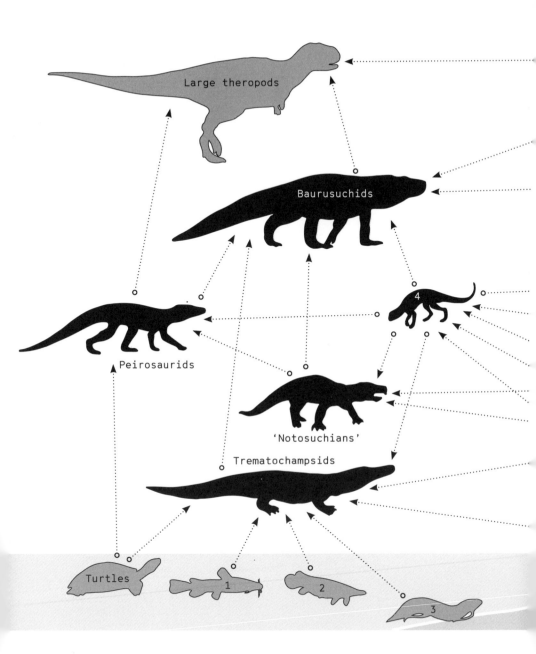

Large theropods

Baurusuchids

Peirosaurids

4

'Notosuchians'

Trematochampsids

Turtles

1

2

3

juveniles and weaker, old specimens. Bizarrely, some of the Adamantina crocodilians were insect-eaters, and one or two even specialized on a diet of plants. When these plant-eaters were first identified, people doubted that such weird crocodilians would have been possible – but the teeth do not lie. There were nearly twenty species of crocodiles in the Adamantina, and this emphasizes how different the faunas of the past were – indeed,

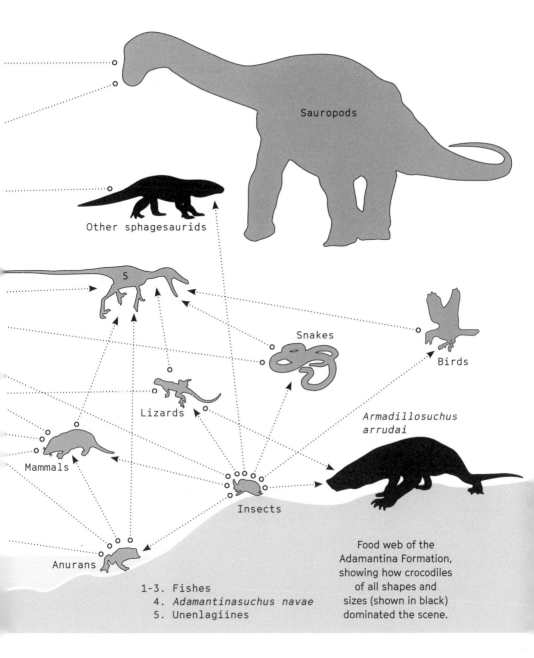

Sauropods

Other sphagesaurids

Snakes

Birds

Lizards

Armadillosuchus arrudai

Mammals

Insects

Anurans

1-3. Fishes
4. *Adamantinasuchus navae*
5. Unenlagiines

Food web of the Adamantina Formation, showing how crocodiles of all shapes and sizes (shown in black) dominated the scene.

crocodilians had a much wider role than they now do, and in South America, many survived the end-Cretaceous mass extinction (see Chapter 9) and continued as important predators until flesh-eating mammals outcompeted them.

In the Adamantina fauna, there were nine dinosaurs (three large theropods, one slender theropod, and five sauropods), but their diets cannot be differentiated, so they are bunched together in the diagram. In the future, careful studies of coprolites, tooth marks, or other evidence, such as isotopes, might help differentiate their diets. Isotopes of oxygen and nitrogen in the bones of ancient vertebrates can give evidence about diet, such as whether they were feeding on fish or tetrapods such as lizards or mammals, but such work is subtle – the isotopes are determined also by climatic conditions, and they can be altered by fossilization processes.

The beetle shown in this food web is a dung beetle, and these helpful creatures have been noted before in association with other dinosaur faunas. For example, Karen Chin, introduced earlier as the finder of the giant *T. rex* coprolite, described burrows made by dung beetles in dinosaur faeces from the Two Medicine Formation, in the Late Cretaceous of Montana. The burrows show that the dinosaurs had been eating conifer leaves, and that the dung beetles in turn burrowed through the excrement, feeding on remaining nutritious materials, and then buried it, just as their cousins do today. Chin noted at the end of her 1996 article, 'This find also reveals a pathway through which fecal resources were recycled and suggests that scarabs evolved coprophagy through association with dinosaurs.' Coprophagy is the eating of excrement. Her words proved prophetic, when genomic evidence was presented in 2016 that scarab beetles had indeed evolved by the Early Cretaceous, and they presumably had the time of their lives, delving in the tonnes of poop dropped every day by the burgeoning herbivorous dinosaurs.

Collapsing food webs

The dinosaur food web is a human construct, relying on our knowledge of how food webs work today, and rare observations from the fossils (coprolites, tooth marks, etc). However, once constructed, the food webs can be used in advanced computational studies. For example, there are mathematical methods now to determine the stability of an ecosystem by knocking out individual species and predicting how the food web would recover. Estimates are made of who eats whom, and the approximate biomass of each species (that is, the body mass of the individual multiplied by relative abundance). If an ecosystem is stable, you can knock out several species, and it will continue to function. If it is unstable, say after a major environmental crisis, the removal of a couple of species can cause the whole system to collapse.

As ecosystems became more and more complex through the Cretaceous, with the addition of flowering plants, new insect groups, and new groups of lizards, birds, and mammals to the existing dinosaur communities, they paradoxically became both more robust and more vulnerable. The close interactions between plants and herbivores and prey and predators gave added robustness to sections of the ecosystem, where species could be clipped out and others would evolve to fill their places without the whole system breaking down. However, overall, with a highly complex ecosystem, as exists today, and existed in the Cretaceous, a major environmental crisis could cause the whole structure to collapse like a house of cards.

New mathematical modelling tools allow ecologists to explore risk and vulnerability in modern natural systems, with an eye on human threats and conservation of course. Likewise, these methods can be applied to fossil examples, both before and after mass extinctions. An initial study by Peter Roopnarine and colleagues has shown how the very last dinosaur-dominated communities of the latest Cretaceous of North America had a so-called 'lower collapse threshold' than those that had existed before, meaning it took less of an environmental shove to cause them to collapse. This seems to have been related to two changes in the last 10 million years of dinosaur ecosystems, with an increase in local forms (those found only in a single locality or small region), and the loss of a number of larger herbivores, which sat at the centre of many connections in the food web. Without them, the latest Cretaceous ecosystems were more vulnerable than those that went before.

Niche division and specialization in feeding

A problem with the Adamantina food web was to discriminate among the five species of sauropods – did they all eat the same kinds of plant food, or did they somehow partition resources? This question has not yet been answered for the Adamantina ecosystem, but such studies have shed light on the diversity of herbivores in the Morrison Formation, as we shall see.

In modern ecosystems, animals typically specialize on parts of the available food resources, perhaps feeding on grass, leaves, fruits, or nuts, or perhaps feeding at different levels – close to the ground, at mid-height, or at tree level. Such specialization has advantages, so that each species evolves specific teeth or digestive systems to best cope with their particular food, and they avoid competition. Just as competition between individuals and between species is core to evolution and to ecology, avoiding competition is a normal response.

An interesting dinosaurian conundrum has been to understand the unusually high diversity of sauropods in the Morrison Formation of the American Midwest, home of the top predators *Allosaurus* and *Ceratosaurus*, and the plated herbivore **Stegosaurus**. Here, ten sauropods (*Amphicoelias, Apatosaurus, Barosaurus, Brachiosaurus,* **Camarasaurus**

Genus:	***Stegosaurus***
Species:	*stenops*

(see overleaf), **Diplodocus** (see pp. 210–11), *Haplocanthosaurus*, *Kaateodocus*, *Supersaurus* and *Suuwassea*) have been found and, although the Morrison Formation rocks span a period of about 10 million years, up to five of these sauropods shared the landscape at any one time. How are we to understand that they could live side by side? The assumption is that there was some division of labour or specialization, and neck length was noted as a clue. For example, *Brachiosaurus* had a very long neck

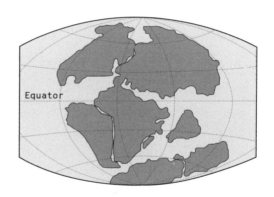

Equator

Named by:	**Othniel Marsh, 1887**
Age:	**Late Jurassic, 157–152 million years ago**
Fossil location:	**United States, Tanzania, Portugal**
Classification:	**Dinosauria: Ornithischia: Thyreophora: Stegosauria**
Length:	**9 m (30 ft)**
Weight:	**4.7 tonnes (10,364 lbs)**
Little-known fact:	**The tail spikes were used in defence, proved by a puncture in an *Allosaurus* vertebra that fits the spike perfectly.**

Genus:	**_Camarasaurus_**
Species:	_supremus_

Named by:	**Edward Cope, 1877**
Age:	**Late Jurassic, 157–152 million years ago**
Fossil location:	**United States, Tanzania**
Classification:	**Dinosauria: Saurischia: Sauropodomorpha: Camarasauridae**
Length:	**15 m (49 ft)**
Weight:	**18 tonnes (36,683 lbs)**
Little-known fact:	**One specimen of a pelvis of _Camarasaurus_ shows evidence of gouging by the teeth of top predator _Allosaurus_.**

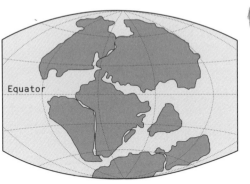

and long forelimbs, so it likely could raise its head, giraffe-like, to feed at great heights, whereas the equally long-necked _Diplodocus_ was more horizontally organized, and so probably fed on leaves at lower levels.

In a functional study of this problem, David Button, one of Emily Rayfield's team, explored jaw function in the two sauropods _Camarasaurus_ and _Diplodocus_. He carried out FEA of their jaws and skulls, and found that _Camarasaurus_ could exert and accommodate greater bite forces than

Diplodocus, suggesting that it ate harder food items. This is supported by his analysis of stresses in the skull, in which *Camarasaurus* showed lower stress values than *Diplodocus*, confirming that the skull of the first was able to withstand larger forces (see pl. ix). Extending his study to a wider sample of thirty-five sauropod species, he found that he could classify each species into one or other of these functional categories according to their estimated bite force, robustness of the skull, and how

Genus:	**Diplodocus**
Species:	*carnegii*

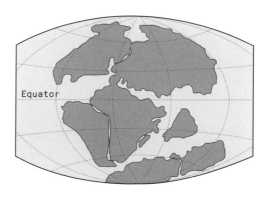

Equator

Named by:	**John Hatcher, 1901**
Age:	**Late Jurassic, 157–152 million years ago**
Fossil location:	**United States, Tanzania**
Classification:	**Dinosauria: Saurischia: Sauropodomorpha: Diplodocidae**
Length:	**25 m (82 ft)**
Weight:	**16 tonnes (35,274 lbs)**
Little-known fact:	**The species is named after multimillionaire Andrew Carnegie, who funded the excavations that found the first specimen.**

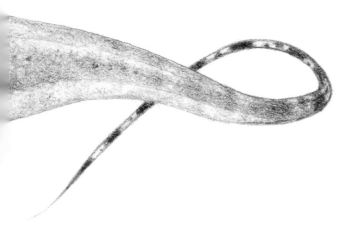

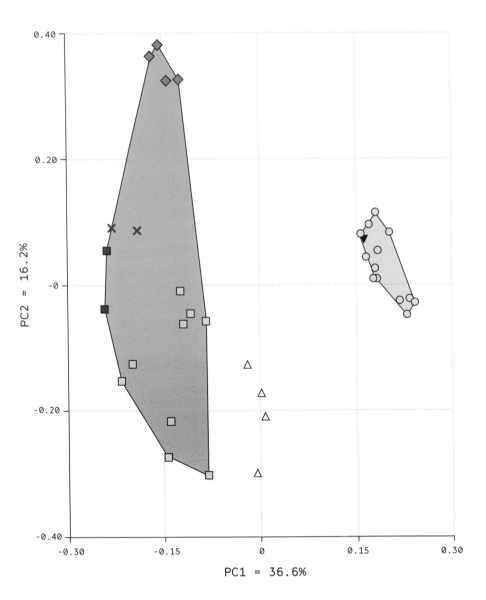

Morphospace plot of different sauropods, reflecting
the feeding habits of branch strippers (left) and
chompers (right).

the teeth interacted with each other. In his morphospace diagram, the two functional classes are spaced a long way apart, and the *Camarasaurus* species and their functional analogues occupy a remarkably tight area, indicating considerable similarities.

Button concluded that *Camarasaurus* was probably a generalized browser that fed on hard and even woody material, while *Diplodocus* specialized on softer, but abrasive, plant materials such as horsetails and ferns. *Diplodocus* has a goofy expression, with a cluster of forward-pointing pencil-shaped teeth concentrated at the front of its jaws, and the researchers posited that it might have been a branch-stripper, gripping a leafy branch in its mouth, and pulling back with the teeth clenched tightly shut, swallowing the leaves as they were stripped. *Diplodocus* had muscles at the back of the head and along the neck that might have enabled this strong backwards pull and twist. *Camarasaurus* had heavier, shorter teeth along the whole length of its jaws, and so it probably used a more normal approach to acquiring plant material by snipping and chomping bunches of branches and leaves, without separating leaves and woody material.

....

We have come a long way in our understanding of dinosaur feeding in the past twenty years. Emily Rayfield summarizes her take on this:

> When I started my research, we had some information from the fossils, such as tooth shape, tooth marks, stomach stones, and coprolites. A few experts in biomechanics had suggested ways to model dinosaur jaws like levers, so you could make some basic calculations, but we now have integrated computational methods that allow much more complex – or realistic – questions to be asked.

In a 2018 TV programme about ichthyosaurs, the dolphin-shaped marine reptiles, David Attenborough, the host, asked Emily her opinion: 'So this was the king of the Jurassic sea?' 'Or queen,' came back Emily in a flash.

The new engineering approaches are all testable, so palaeontologists are no longer speculating about feeding in extinct animals. Smart new approaches in ecology, especially using food webs, are also beginning to help, but there is so much more to do. We can expect to see an integration of both approaches soon, with dinosaur food webs modelled with accurate data on feeding mechanics and diets on the one hand,

and understanding of how robust these ecosystems are to outside environmental pressures on the other. Emily Rayfield's main intention now is to work hard with her team in another direction, to integrate our understanding of feeding modes and styles with evolution, so we can document why some dinosaur groups were more successful than others, and how different adaptations for feeding might have fared through the Mesozoic.

Chapter 8

How Did They Move and Run?

The study of dinosaur locomotion is a perfect example of how palaeobiology has shifted from speculation to science. Two pioneers drove this revolution, one an inspirational English professor with a long beard, the other an American professor who settled in England, but so far without sprouting a beard.

The first professor is the legendary late biomechanics expert, R. McNeill Alexander (1934–2016) from the University of Leeds. He led the field, as the author of papers on the biomechanics of everything from fishes to mammals, as well as numerous standard textbooks such as his *Animal Mechanics*, *Functional Design in Fishes*, and *Locomotion of Animals*, published through the 1960s to 2000s. His lectures were famous: with his extensive knowledge of the function of all animals from fleas to elephants, and with his lanky frame and flowing beard, he would mimic animals hopping, jumping, and flying. Alexander's lilting Ulster accent, retained from his birth and schooling in Northern Ireland, added engaging colour to his speech. From time to time, Alexander made a foray into the world of dinosaurian palaeobiology, with suggestions about how to estimate their body mass using plastic models or how to calculate their running speed. His insights on dinosaur running speed changed the field; overnight, in 1976, palaeontologists were provided with a reliable formula that told them the speed, not just a guess.

John Hutchinson came to the field in a newer generation, but he had been influenced by McNeill Alexander's books. 'Alexander wrote like a dream, and his insights were so clear you realized there was no limit to how we could explore the locomotion of modern animals and apply the findings to dinosaurs,' says Hutchinson. Hutchinson is a stockier individual than Alexander, shaven-headed and built like an American football player, but gentle and madly enthusiastic, just as Alexander was. Hutchinson is famed for his TV shows in which he dissects an elephant, a racehorse, or an ostrich. He writes a blog called 'What's in John's Freezer', and he was characterized in a 2011 National Geographic blog as the man

Professor R. McNeill Alexander explaining how to estimate the original mass of a dinosaur.

John Hutchinson, in unusually serious mood, surrounded by bones.

who had the most frozen elephants' feet. Working as he does at the Royal Veterinary College north of London, he has access to exotic animals that arrive for investigation, many from local zoos. One week he may be studying whether elephants can run (the books say they can't; John's videos show they can, and in a very unusual way), and the next advising Hollywood dino-film producers, or helping students with their computer code in locomotion mechanics.

John Hutchinson's PhD was a classic anatomical study of the muscles of the legs and hips of birds and crocodiles:

> This gave me the groundwork to study dinosaurs. I dissected nine alligators, as well as lizards, snakes, turtles, and dozens of birds, and worked out the common shared patterns of all the muscles that power their locomotion [John says]. When I began my PhD, I was interested in how organisms have evolved. I wasn't wedded entirely to one idea but had enjoyed the novel and movie of *Jurassic Park* and saw an opportunity to apply modern biomechanics methods to test how *T. rex* might have stood and moved. I ended up sticking with that plan throughout my PhD and fortunately it worked out pretty well. Fixing the basics of musculature, that it's much the same across all vertebrates, was helpful.

This is the idea of the extant phylogenetic bracket (see p. 17), which allows palaeontologists to determine most details of the leg muscles of dinosaurs with certainty based on their bracketing descendants, the crocodiles and the birds. As Hutchinson found in his dissections, all vertebrates have the same basic limb muscles, so it's most likely the muscles of dinosaurs were the same too. The palaeontologist then simply has to determine the sizes of those muscles by looking for muscle scars on the bones, and these attachment sites can often be well defined and give a good measure of how broad a muscle was. The width or diameter of a muscle gives its strength, and the positioning of the attachment sites shows its direction, and can be used to calculate forces. But, as John explains:

> ...it's more than just the muscles. We need to understand how an animal holds its legs, how its body mass is distributed through the legs to the ground, and the sequence of movements of the joints and firing of the muscles. There are also all the different ways of moving, the gaits. We know a horse has

five gaits, the walk, the pace, the trot, the canter, and the gallop. Now, do all animals show these same gaits? What about dinosaurs?

Understanding how dinosaurs moved involves a remarkable mix of fashion and science. Until recently, the way we viewed dinosaurs – whether as lumbering, slow reptiles, or active, fast movers – depended on the prejudices of the moment. Evidence from footprints and skeletons forms the basis, and the new computational methods have allowed researchers like McNeill Alexander and John Hutchinson to test what is likely and what is unlikely. Now we know for sure how dinosaurs stood and walked, their speeds, and whether they could fly and swim. We can also then compare the science with the Hollywood glitz – did they get it right in the movies?

Fashions in dinosaur posture and locomotion

There have been fashions in how we reconstruct dinosaurs. How we imagine dinosaurs has evolved from giant crocodile to giant rhinoceros to kangaroo to balanced high-speed biped, to somewhat more stately biped today. It's all about interpreting the skeleton, finding a plausible posture for the limbs, and choosing the appropriate modern examples. We may treat the older images with humour – how could they ever have believed that? – but palaeontologists of the future will likely mock our best efforts. Nevertheless, we might still hope that our hypotheses improve through time: 'we stand on the shoulders' of previous researchers, as Isaac Newton said.

When dinosaurs were first reconstructed, palaeontologists thought they were looking at outsized lizards or crocodiles. About 1830, Gideon Mantell even reconstructed his *Iguanodon* as an enormous lizard, over 61 metres (200 feet) long, walking on all fours and with its body close to the ground. Other dinosaurs were seen then as huge crocodiles, also quadrupedal and also low-slung, but presumably sluggish in pursuit of their equally sluggish prey. As we saw earlier in this book, Richard Owen made a radical revision to the image of the dinosaur in the 1840s and 1850s, picturing them as warm-blooded, rhinoceros-like animals, portly and presumably slow-moving, but at the same time warm-blooded.

Owen's rhino-dino did not last long, because complete skeletons from North America showed he had got it wrong. The first hadrosaur skeleton

An early vision of *Iguanodon*, reconstructed as a
61-metre-long (200-ft) lizard by Gideon Mantell.

Richard Owen's massive, almost rhinoceros-like,
vision of *Iguanodon*, from 1853.

reported by Joseph Leidy in 1858 was clearly a bipedal animal, whose
legs were three or four times as massive as its arms. Leidy opted for a
rather vertical posture, in which the animal sat back on its heels and the
tail touched the ground; its torso then stood vertical, like that of an alert
kangaroo. This standing posture prevailed until 1970. Other images of
bipedal dinosaurs after 1860 were equivocal about the running posture

of bipedal dinosaurs – some realized the animal should be flipped to horizontal, with the backbone and tail straight out, and the body in front of the hips balancing the tail behind – more see-saw than kangaroo. Others persisted in trying to show the dinosaur running with its body vertical and the tail trailing along the ground – an impossible contrivance that required several elements of the tail and neck to be broken. At speed,

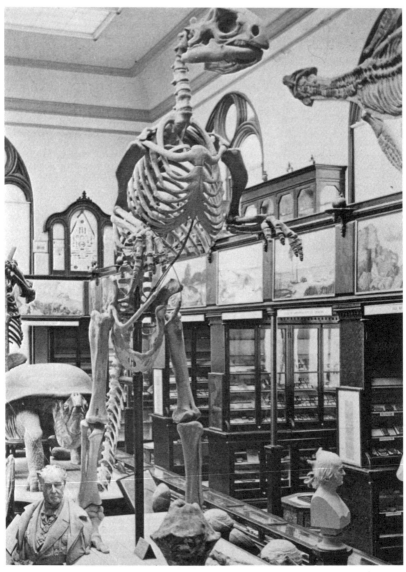

Replica cast of *Hadrosaurus foulkii*, the first complete dinosaur skeleton to be put on display, shown here at Princeton University's Nassau Hall in 1898.

such a dinosaur would have been more Monty Python silly walker than believable moving creature.

The revolution came in 1970, when two young palaeontologists, one British, one American, hit on the same realization. As we saw in Chapter 4, Bob Bakker drew the small theropod *Deinonychus* as a lively, sleek horizontal runner, and Peter Galton did the same for the hadrosaur *Anatosaurus*, shown stretched horizontal and 'in a hurry', as he stated. The truth of their insight was grasped immediately, and the kangaroo-dinosaurs were never to be seen again, except in the cheapest of kids' books. For Bakker and Galton, the modern comparator was not the kangaroo, but the road runner. This may be known to most as the cartoon character who tore around the US desert pursued by Wile E. Coyote, but it is also a real bird, slender and famed for running fast with tail and head outstretched.

The dinosaur skeleton speaks in favour of Bakker's and Galton's insights. First, with a decent skeleton, the nature of the joints can be determined. In fact, dinosaur limb joints are as simple as those of birds and mammals, including humans, today. In the leg the knee and ankle are simple hinges, and in quadrupedal dinosaurs, the same is true of the wrist and elbow in the forelimb. Bipedal dinosaurs, like birds and humans, have arms capable of more complex movements, with rotation at wrist and elbow to allow birds to flap their wings, humans to play tennis, and apes to swing hand over hand through the trees. In all cases, the hip and shoulder joints allow rotation as well as back-and-forwards movements.

Dinosaurs had all adopted an upright or erect posture, which distinguishes them from their sprawling ancestors. At the time of the

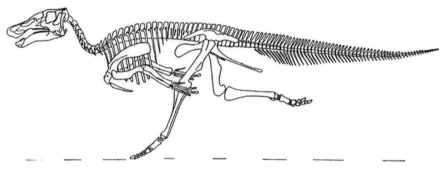

Peter Galton's 1970 reconstruction of *Anatosaurus* in a hurry.

Permian–Triassic mass extinction, 252 million years ago, all the large sprawling animals died out, and the new groups that replaced them in the Triassic (see Chapter 1) adopted an erect gait. Sprawlers today include small animals such as salamanders and lizards, in which the arms and legs stick out to the side, and swing widely as the animal moves. The body is held close to the ground, and sprawlers cannot run fast for very long – the effort of holding the belly up is huge, and the stride length is limited. The new erect animals could strut about fast and exert very little effort holding themselves upright.

Indeed, dinosaurs were fundamentally bipedal from the start, and that would have been impossible if they had been sprawlers. It's not that erect posture evolved to enable bipedalism – the immediate advantage being speed and escape from predators – but bipedalism and huge size, and ultimately flight in advanced theropods and birds, were later enabled by the erect posture. The sprawling posture in lizards means their ankles, wrists, knees, and elbows are much more complex than the hinge joints seen in erect-postured dinosaurs and mammals, as each limb goes through numerous twists and turns during a sprawling stride.

Some of the first evidence about dinosaur locomotion, however, came from footprints, not skeletons.

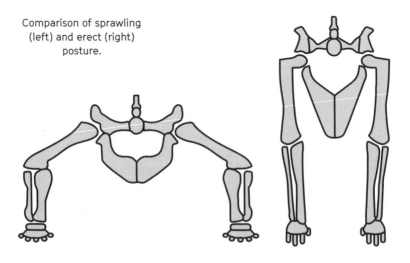

Comparison of sprawling (left) and erect (right) posture.

What can we learn from tracks and trackways?

Dinosaur tracks have been known for more than 200 years, even though early researchers were unsure what they were. The first published record dates from 1807, when Edward Hitchcock, a clergyman based in Amherst, Connecticut, was shown some large three-toed footprints in red sandstones from the local Upper Triassic rocks. The specimens had been observed first by a farm boy, Pliny Moody, in 1802, who, it is said, levered them up and made a doorstep out of the slab.

Hitchcock was hooked. Over the next thirty years, he sought more specimens, and they were readily forthcoming as stonemasons opened quarries around Hartford, Connecticut, seeking red and yellow sandstones for house building. As they levered off the slabs, they saw individual footprints and whole tracks, often containing fifty or more footprints, neatly repeating left-right-left-right across the surface as far as you could trace them. Hitchcock began to collect the slabs and eventually donated them to Amherst College, where they may still be seen. He published several elaborately illustrated monographs describing his finds, most notably his 1858 book, *The Ichnology of New England.*

The first dinosaur footprints to be discovered,
on show in Connecticut about 1810.

One of the Hitchcock dinosaur footprint slabs in the Amherst Museum, Connecticut; the two slabs show the underside (above) and mould (right) of the same set of four 3-toed imprints.

Hitchcock always thought these were bird footprints – and some of them were pretty big birds, admittedly. He could not equate them with any modern reptile such as a crocodile or lizard, which generally have five fingers and toes and put the palms and soles of their hands and feet flat on the ground. The tracks from the Triassic of the Connecticut Valley were three-toed, like modern birds. Theological friends of Hitchcock's speculated whether they were antediluvian, and had perhaps been made by Noah's raven as it sought dry land after Noah's flood. There were so many tracks, though, and many were so huge, that this was hardly likely – unless the raven had been twice the size of an ostrich!

Now we know that these tracks were made by early dinosaurs, animals like the 2-metre-long (6½-foot) *Anchisaurus*, known from its long, slender skeletons in rocks of the same age in New England. *Anchisaurus* had a tiny skull and plant-eating teeth, a long neck and tail, and a slender body supported by arms and legs of similar length – it could probably flip between walking on all fours and running on its hind legs. Larger prints might have been made by relatives like *Massospondylus*, up to 6 metres (20 feet) long. Both these dinosaurs are sauropodomorphs, ancestors of the long-necked *Brontosaurus* and the other sauropods.

Why did these animals leave so many tracks in Connecticut? It's probable that such animals were running around in all parts of the world, but the conditions of rock deposition happened to be just right in the Connecticut Valley, as well as in footprint localities of similar age in South Africa. In New England, the conditions were unusual – hot climates, with great lakes surrounded by vegetation, and rich faunas of fishes in the lakes, small lizard-like animals snapping at cockroaches and dragonflies at the water's edge, and the plant-eating dinosaurs crowding on the shore to find the richest plant food.

The hazy heat of midday in that era reminds us that New England, like Germany and North Africa, lay close to the equator then, and the Atlantic Ocean was just beginning to unzip between Europe and North America. Volcanic eruptions and tearing movements between the future tectonic plates of Europe, North America, and Africa opened deep rifts, just as in the Great Rift Valley of eastern Africa today. The rifts formed natural depressions in which lakes small and large could form, and these were a magnet for insects, plants, and fishes, and for the dinosaurs that fed on these food supplies.

After 1850, dinosaur tracks were found in increasing numbers in the Late Triassic, Jurassic, and Cretaceous worldwide. Palaeontologists became adept at matching tracks to likely makers, based on the size and shape of the toe impressions, whether the toes of three-toed prints were tipped with sharp claws (meaning theropod makers) or not (meaning ornithopod makers), or whether the prints were huge circular markings, presumably made by giant dinosaurs such as sauropods.

As Hitchcock had noted, many of the Connecticut Valley footprints were deeply impressed in the sediment. It must have been very soft, he surmised, as the prints go down as much as 10 or 20 centimetres (4 or 8 inches), penetrating through numerous layers of rock. He could not put this amazing data to use, but locomotion experts Stephen Gatesy and Peter Falkingham were able to apply scanning and digital engineering technology to the specimens. They reconstructed the foot cycle to show the slender left foot entering the silt with all three main toes extended as broadly as possible, sinking down through the sludge until it hit firmer silt or sand, and then taking the animal's weight as its body continued forwards, and the leg followed. Then, as the right foot grounded and took the weight, it pulled the left foot up, but with the toes now bunched tightly together and curled like an arthritic witch's hand so they could pull free of the gloop with minimal effort.

This entire foot cycle can be seen in the multiple layers of sediment, and in the CT scans and models – the foot goes in with toes out, grounds on firm sediment, rolls forwards as the animal keeps moving, bunches the toes tightly and pulls out, as the other foot begins its next stride.

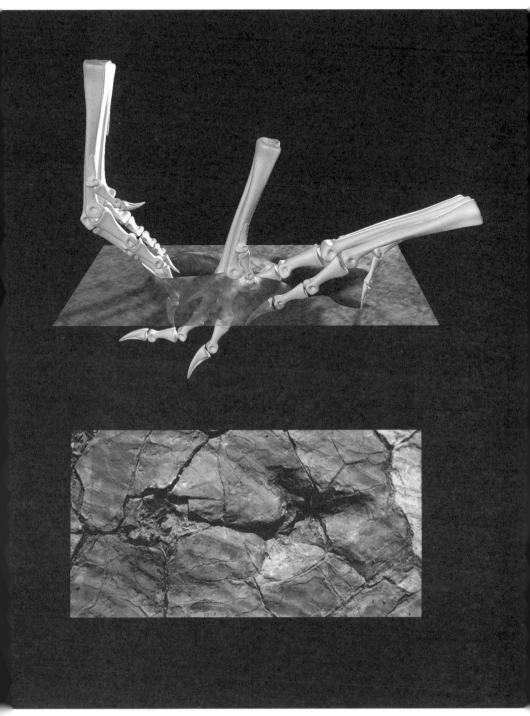

The slender dinosaur foot enters the sediment with
toes splayed, and then withdraws, as the animal
moves forward, with the toes clenched.

How fast did dinosaurs run?

It might seem impossible to work out how fast dinosaurs walked and ran, but in fact this was one of the most basic of calculations. The method was pointed out in 1976 by McNeill Alexander. He had noticed that there was a 'rule' of locomotion, which seemed to hold for bipeds and quadrupeds. It was based on a commonplace observation that as speed increases, so too does stride length – when speeding along very fast, the stride length is twice or three times the length of the walking stride.

McNeill Alexander knew from basic biomechanics that all animals should show the same relationship between speed and size – in other words, as they become larger, they should move relatively slowly. His inspiration came from a fundamental formula of shipbuilding, of all things. In 1861, the great Victorian engineer William Froude established

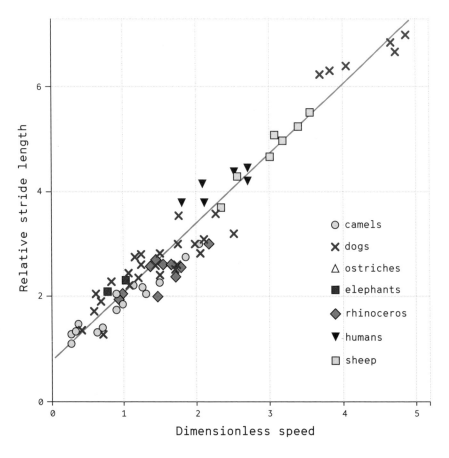

The relationship between relative stride length and dimensionless speed.

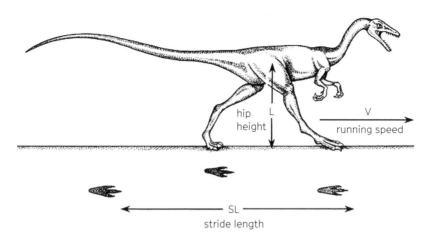

Estimating speed from stride length and hip height.

a fundamental law of mechanics that became essential for ship design: he noted that the resistance in a fluid to the motion of a ship is proportional to the square of its speed relative to the fluid and the area of resistance. McNeill Alexander was the first to realize that this formula could apply to a whale as well as to a ship, but it could also be modified to refer to running as much as to swimming. He found that he could describe the walking and running of a wide variety of living animals as Froude number = 2.3 (relative stride)$^{0.3}$. The speed proportional to body size is the so-called 'dimensionless speed', or simply 'relative speed'.

This was the basis for McNeill Alexander's famous paper of 1976, only two pages in length, but in which he made the proposal that we could calculate dinosaur running speed with confidence from a trackway. His argument was undeniable – we have a formula[1] that always works for modern animals. He tried it out on humans, horses, dogs, elephants, birds, rats...it always worked. I remember seeing his wonderful demonstration in a BBC Horizon programme back in the 1970s, when he and his somewhat put-upon family interrupted their holiday to North Norfolk to run along the beach, and then repeat with dogs and horses, while the good professor, vast untrimmed beard flying in the wind, measured their stride lengths with a ruler.

McNeill Alexander suggested his formula could be applied to dinosaurs. Here was a simple way to estimate running speed from a trackway – and palaeontologists had thousands of tracks to look at.

----

1 $v = 0.25\,g^{0.5} \times SL^{1.67} \times h^{-1.17}$, where '$v$' is velocity (speed), '$h$' is the height of the hip, 'SL' is stride length, and 'g' is gravity (10 metres per second per second).

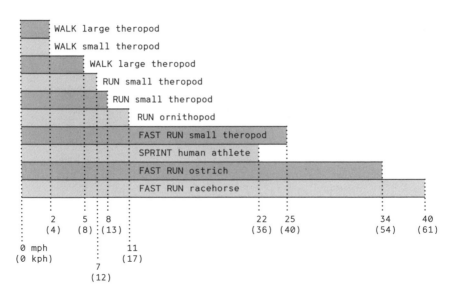

Walking and running speeds of dinosaurs,
compared to some modern animals.

Typical speeds were in the range of 1–3.6 metres per second (equivalent to about 2 to 8 miles per hour), and the debate became heated as palaeontologists like Bob Bakker used his estimates of faster speeds, such as 20 metres per second (45 mph) for *Tyrannosaurus*, to demonstrate that certain dinosaurs ran fast and so must have been warm-blooded, and argued, as did others, that we would never be able to use tracks to indicate maximum speeds, because nobody can run at their fastest on wet mud or sand. Fast running requires a hard surface, and this is just the kind of surface that does not preserve tracks.

Nevertheless, most calculated dinosaur speeds are modest. Much attention has focused on *T. rex*, but it has been hard to marry tracks directly to the adult animal. Juvenile tracks found in 2016 suggest a speed of between 2.8 and 5 miles per hour (1.3 and 2.2 metres per second), a brisk walk for a human, but pretty slow for a larger animal. So, life in the Mesozoic may have proceeded in slow motion, not at the speed of race horses and cheetahs. This is a maximum possible speed estimate for *T. rex*, and tracks provide matching speed estimates.

How do these speed estimates fit more detailed modelling of locomotion? Digital modelling must obey the laws of physics, of course, but also fit in with the observational data from fossil trackways. John Hutchinson, like all other biomechanics experts, had to work from first principles. He recalls, as he worked on his doctoral thesis:

the dissecting and museum visits were a ton of work. Once I finished, I then had to dive into the physics of locomotion, which was intimidating especially considering how much was unknown about *T. rex*. I had to devise a way to take these unknowns into account while still testing how *T. rex* could have moved. At times, I gave up hope for finding anything!

Digitizing dinosaurs: how did their legs work?

Hutchinson did not give up. He knew that engineers use a great deal of common sense in their work, but they are also thorough. When he and collaborator Stephen Gatesy thought about the true posture of dinosaurs, they decided they should try out every possibility, and chose *Tyrannosaurus rex* as the model. So, they generated an amazing array of postures – thousands of them (see selection overleaf). Some looked quite sensible, with the legs pacing along at mid-height; but they also tried the Russian Cossack walk, in which the animal was squatting down close to the ground and somehow propelling itself forwards with its knees round its ears, then the fairy twinkle-toes look, with the body as high as it could go, and the legs tripping along almost *en pointe*.

These extremes could be ruled out immediately, because they involved huge extra forces to keep the legs ramrod straight or in a crouching posture. The so-called 'ground reaction force' is the key – this is the force acting vertically upwards as the equal and opposite force to the body weight. If it passes vertically upwards in front of the knee, such as when the limb is held too straight, the animal falls over backwards; if it is too far behind the knee, the rotational force about the knee is too high and could not be resisted by any reasonable size of leg muscles. In this case, two postures ('a', 'b') can be ruled out, whereas 'c' might be feasible.

After numerous runs of the experiment, Hutchinson and Gatesy settled on a core set of postures and strides, and these were the ones most people would have agreed looked more or less plausible. The experiment, however, was helpful as it gave logical reasons, in terms of energy wastage, for why the others should be rejected. The question then was whether such a pure-biomechanics approach could be used to determine maximum running speeds from first principles. We will never be able to follow a running *T. rex* and time its speed, but if the speed calculations from footprints on the one hand, and the fundamentals of skeletal

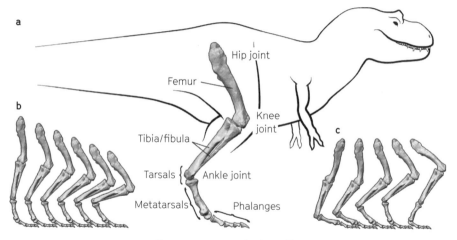

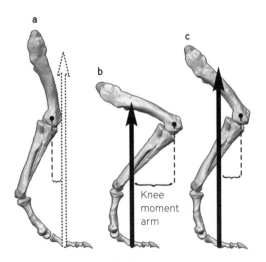

Tyrannosaurus rex body outline,
and various possible leg postures.

The knee moment arm is the rotational force around
the knee, and posture (c) shows the correct position
of the knee to minimize the force. In 'a' the leg is too
straight and in 'b' it is too crouched.

function on the other, give the same answer, then we might suggest this is correct. It's not a perfect way of doing science, but it satisfies the demands of common sense – and can be said to constitute proof, in a legal sense at least.

Hutchinson's idea, in joint work published in 2002 with Mariano Garcia, was to use their knowledge of skeletons and muscles to estimate speeds. There is a standard relationship between muscle volume and speed, based on the assumption that the force of a muscle is proportional to its cross section. We see this when we compare the legs of a sprinter and a non-athletic person – the sprinter's extensor muscles, which provide the main power in running, may be twice the diameter of those of a normal, healthy, but not obese person. Likewise, with other animals – greyhounds and whippets are all muscle when compared to other dogs.

The muscle volume versus force or speed relationship scales with body mass. Smaller animals need relatively less muscle to achieve fast speeds (which themselves are proportional to body size). Hutchinson took the chicken as his example of a fast-running small animal. This might not seem a good choice, but chickens are built for speed on the ground, as anyone who has tried to catch one can attest. A chicken weighs about 1 kilogram (2 pounds), and Hutchinson calculated its main leg muscles

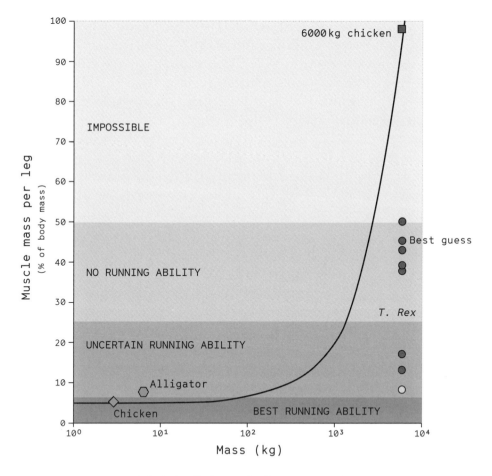

Calculating likely speeds of movement and
the muscle mass needed to achieve that
speed, at increasing body mass.

comprised 10 per cent of that body mass. He then calculated the volume of leg musculature required by a 6-tonne chicken (the size of *T. rex*), and the answer was that it would have needed about 10-tonne leg muscles, that is 200 per cent of the animal's body mass (100 per cent × 2 legs = 200 per cent), to power it at an equivalent speed to a charging chicken. This is not possible – no animal can devote twice its total body weight to leg muscles. Not even 50 per cent would be possible, because all the other organs are required to make a functioning animal.

In a famous image produced to highlight their work, Hutchinson and Garcia showed the 6-tonne chicken keeping pace with a *T. rex*. Their calculation was that *T. rex* could have devoted no more than 30 per cent of its body mass of 6 tonnes to the leg muscles, and this meant it scales as a gentle stroller, capable of speeds of, at most, 10–22 miles per hour. These are the absolute maxima, and speeds of half these values are more likely. Recall that the footprint data gave a speed of 2.8–5 miles per hour. This represents good independent corroboration based on two independent scientific observations – fossilized trackways on the one hand, and on the other a well-established rule of biomechanics concerning body size scaling and the relative size of the leg muscles.

Recent work by John Hutchinson and his group shows how much there is yet to learn. In a 2018 study of modern birds, they confirmed a formerly debated point, that in fact most modern birds can move from one gait to another – say, walking to running – in a smooth manner, simply by moving faster, whereas in humans and many other mammals, there are distinct switches from walk to run. Applied to *T. rex*, the new bird locomotion model predicts that it would have bounced along at a steady lope, making strides 4 metres (13 feet) long, but always with at least one foot on the ground (it did not have a so-called airborne phase, as seen in fast runners such as ostriches and racehorses).

These recent studies illustrate a new level of confidence in a field that McNeill Alexander pioneered forty years ago. According to John Hutchinson:

> We can now do much better in animating dinosaur locomotion than just using intuition. Evidence from footprints, skeletons, biomechanics, and comparisons with modern forms allows us to test likely poses and likely gaits. We can even identify cases – as with many dinosaurs – where they were moving in ways no modern animal does.

How did *T. rex* use its arms and legs?

The legs of tyrannosaurs were clearly used primarily to support the huge body weight and for locomotion. Perhaps they were also used to hold down the prey. Today, vultures and other scavenging birds use their feet to hold a carcass steady while they rip at the flesh. The African secretary bird, so called because it has a long quill-like feather tucked behind each ear, chases lizards and snakes, traps them by slapping its foot down hard on top of its prey, and then killing and tearing with its beak. Owls and eagles do the same thing, snatching their prey in powerful foot claws and carrying it off, and then holding the prey animal down with the foot to stop it squirming as they tear off flesh.

Birds have to use their feet for holding and subduing prey because they cannot grasp things with their wings. Dinosaurs, however, had freed their arms by being bipedal, and surely the early theropods would have grasped their prey with their hands while they bit at it. Likewise, early bipedal herbivores such as the basal sauropodomorphs and ornithopods were able to grab at and clutch leaves in their hands. Most later plant-

A modern secretary bird shows no fear in quelling a cobra.

Genus:	*Tyrannosaurus*
Species:	*rex*

eaters became quadrupedal, so the arms became pillar-like limbs, and the fingers shortened and were equipped with small hooves. Their arms were no longer capable of grasping. These sauropods, ceratopsians, stegosaurs, ankylosaurs, and hadrosaurs must then have manipulated their plant food almost exclusively using their mouths, having given up their ability to use their arms for anything other than walking.

Most theropods reduced the number of fingers from five to four, three, or even two, in some species of *Tyrannosaurus*. At the same time, the overall size of the arms became reduced until, in *Tyrannosaurus*, the arms were 20 per cent of the length of the legs, compared to 50 per cent in an early theropod such as *Coelophysis*, or 70 per cent in humans. What on Earth were these arms used for? As noted in Chapter 6, this trend for

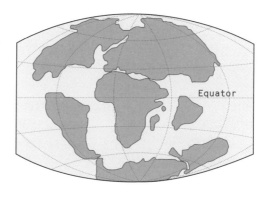

Equator

Named by:	**Henry Osborn, 1905**
Age:	**Late Cretaceous, 68–66 million years ago**
Fossil location:	**United States, Canada**
Classification:	**Dinosauria: Saurischia: Tyrannosauridae**
Length:	**12.3 m (40¼ ft)**
Weight:	**7.7 tonnes (16,978 lbs)**
Little-known fact:	**One *T. rex* specimen, nicknamed Sue after its discoverer, Sue Hendrickson, was sold to the Field Museum, Chicago for $8.36 million in 1997, the highest-priced dinosaur ever.**

The huge size of *Tyrannosaurus rex* can be appreciated with a human for scale.

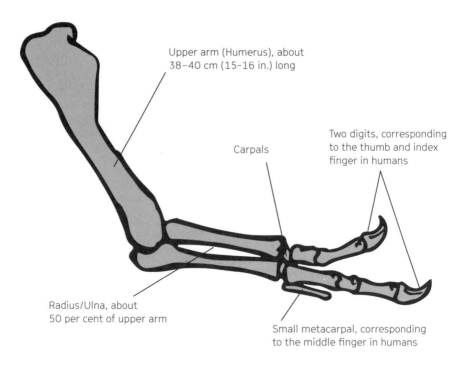

Upper arm (Humerus), about
38–40 cm (15-16 in.) long

Carpals

Two digits, corresponding
to the thumb and index
finger in humans

Radius/Ulna, about
50 per cent of upper arm

Small metacarpal, corresponding
to the middle finger in humans

The tiny, but powerful, arm of *T. rex*.

reduction in the arms was reversed in the maniraptorans, who used their elongate arms in flight. But *T. rex* was stuck with short arms, and (like the legs) they even became relatively smaller as the animal grew from juvenile to adult, as John Hutchinson and colleagues showed in a 2011 study.

The dinky little arms of *T. rex* have been the subject of much speculation. If *T. rex* held down a dead carcass, or killed its prey, with its broad, spreading feet, were the arms used in any way in hunting? The common agreement is that they were not, because they could not reach the mouth – so even if *T. rex* grabbed a tasty morsel in its hand or hands, it could not even shove it into its mouth. Other suggestions are that the arms were used to push the animal up off the ground after it had been asleep, to hold down prey while the death bite was delivered, or even to tickle members of the opposite sex to encourage them to mate. None of these ideas is testable, but studies of the lever mechanics of the arms show that they were strong, even if ridiculously small. Their function remains a mystery, one of those puzzles in dinosaur science that will keep future researchers happily engaged.

It might have been a different matter for another huge theropod with short arms – the unrelated ***Carnotaurus*** from the Late Cretaceous

Genus:	*Carnotaurus*
Species:	*sastrei*

Named by:	José Bonaparte, 1985
Age:	Late Cretaceous, 72–69 million years ago
Fossil location:	Argentina
Classification:	Dinosauria: Saurischia: Abelisauridae
Length:	9 m (30 ft)
Weight:	1.6 tonnes (3,528 lbs)
Little-known fact:	*Carnotaurus* had a pair of horns located on top of its head, and these might have been used by males for head-butting contests.

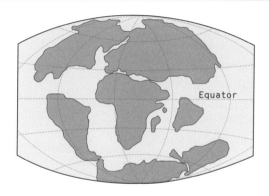

of Argentina. This dinosaur went even further, and had arms only about 12 per cent the length of the hindlimbs, and the wrist bones were much reduced. In fact, this forelimb has been described as 'vestigial', meaning it's barely there – so reduced in size as to be almost without function, like the wings of the flightless emus and kiwis today. Maybe *Carnotaurus* had dispensed with using its arms and hands for grasping, but used them instead to twirl a tuft of feathers on each side – like a Cretaceous exotic dancer, seeking to attract members of the opposite sex.

Would the arms even have eventually disappeared if the tyrannosaurs and carnotaurines had not become extinct at the end of the Cretaceous?

Could dinosaurs swim?

All animals can swim, even cats. They may not like it, but they do it when they have to. Therefore, there is no reason that dinosaurs could not have been swimmers, especially when they were trekking over long distances. It's not known whether all dinosaurs migrated, but it seems likely from our knowledge of modern large mammals. Today, caribou and elephants, for example, are famous for their long migrations as they seek sufficient food supplies in the face of seasonally varying availability.

In the Cretaceous, North America was divided into two land masses, one to the east and one to the west of the Western Interior Seaway, which ran up through Mexico and Texas to Alberta and Northwest Territories.

Mass sets of dinosaur tracks at Dinosaur Ridge, Colorado – mostly heading in the same direction.

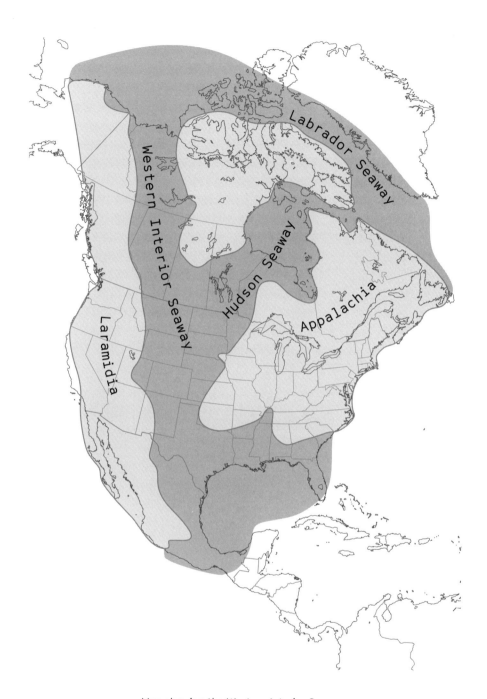

Map showing the Western Interior Seaway.
Dinosaurs trekked north and south along the
east coast of Laramidia.

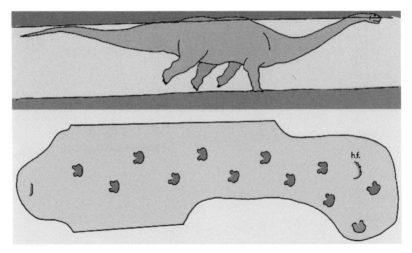

The most likely explanation of the hand-only sauropod trackways.

Martin Lockley, famed dinosaur track enthusiast, born in England but a long-time resident of Colorado, identified a number of what he called dinosaur megatracksites, locations with thousands of footprints, mostly in the form of trackways, on the western coastline of this inland sea. The megatracksites documented how herds of dinosaurs trekked north and south, perhaps covering 2,000–3,000 kilometres (around 1,250–1,850 miles) in a season, in search of lush vegetation. Presumably, during these migrations, the herds had to swim across rivers, just as caribou and wildebeest do today. We can only imagine the immensity of the herd as it passed, the adults, weighing up to 50 tonnes each, clapping their great feet, each as broad as a tree trunk, thunderously to the ground and stirring up great clouds of dust. The juveniles, some as tiny as a sheepdog, would stay in the middle of the moving herd for safety, but could shoot in and out between the legs of their parents.

Some rare tracks appear to support the swimming idea. One of the most peculiar was discovered by buccaneering dinosaur collector Roland T. Bird, and reported in 1944. He had found a series of large, hand-only sauropod prints in the Cretaceous of Texas, and speculated that the animal was moving through deep water, its hind quarters and tail floating, and using its hands to paddle or prod its way through the water, maybe to change direction, as the hindlimbs did a doggy paddle behind. The only other explanation could be that the sauropod was balancing on its hands, and doing something extremely acrobatic. Such a suggestion, amusing as it might be, would be impossible because of the huge weight of the back half of the animal's body. Further, even if the dinosaur were an amazing

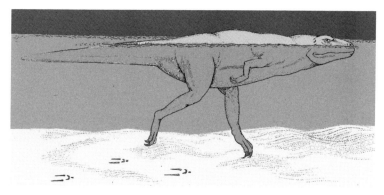

A determined theropod swims in deep water, leaving the merest of scratch marks on the bed of the river.

gymnast, with all its weight expressed through the arms, these would surely have made deep prints in the sediment, not the light prods that are seen in the trackway. Such hand-only sauropod prints have been reported also from South Korea and China, and may represent a regular behaviour.

Roland Bird worked as a collector for the American Museum of Natural History (AMNH), and he discovered sauropod footprints on the banks of the Paluxy River, near Glen Rose, in Texas. He heard about the site from local informants who said the farmers there were busy excavating human footprints from these ancient rocks and selling them to gullible visitors. Here was proof that humans and dinosaurs lived side by side! Bird devoted great efforts to photographing, mapping, and excavating Paluxy River tracks for the AMNH, and did his best to explain to the farmers what they were. The supposed 'human' footprints were just chance bits and pieces of footprints, often a single toe of a three-toed dinosaur print, and yet these continued to be cited as evidence for 'creation science' until quite recently.

Swimming tracks have also been reported for theropods, including scrape marks made by the feet as a *Megalosaurus*-like animal floated along in the Early Jurassic of Connecticut, dabbing occasionally at the bottom of a river with its extended toes to keep moving in the right direction. Debra Mickelson reported a series of theropod tracks from the Late Jurassic of Wyoming that first showed normal tracks as the ostrich-sized animal walked in shallow water, then lighter foot impressions as it moved to ever-deeper water and its body began to float, taking the weight off its feet, and then mere toe-tip scrapes when it finally launched into swimming.

Dinosaurs may not have been especially adapted for swimming, as crocodiles or seals are. For example, they generally did not have deep, narrow tails for beating in the water, or fused fingers and toes, and paddle-like limbs. Still, as already noted, nearly all animals today can swim, even if they show no special adaptations for it – think of horses and cattle swimming strongly enough to cross fast-flowing rivers, just by beating their spindly legs underwater.

Could dinosaurs fly?

Birds can fly, and so too could the pterosaurs, the flying reptiles that were close relatives of dinosaurs. Until the 1990s, most palaeontologists would have said that dinosaurs proper did not fly. However, the discoveries of feathered dinosaurs from China changed everything, as we saw in Chapter 4. In fact, John Ostrom had enough evidence for dinosaur flight back in 1969, when he described *Deinonychus*, but he could not say so. He wondered at the long arms of this dinosaur, and interpreted them as perhaps necessary for predatory behaviours, such as grabbing or wrestling with prey. He and Bob Bakker even speculated whether

A diversity of Maniraptora, or a 'maniraptoran montage'.

Deinonychus might have had feathers along its strong arms, but feathers were not preserved, and other palaeontologists derided such speculations.

In 1986, Jacques Gauthier named this key group of theropods – *Deinonychus*, birds, and their close relatives – as Maniraptora, a wonderful choice of name that means 'hand hunters'. He noted that *Deinonychus* and its kin shared with birds their long arms and other characters, and that in this regard they entirely bucked the general trend in theropod evolution, which saw the arms reducing in size through time – reaching extremes of uselessness, as we have seen, in *T. rex* and *Carnotaurus*.

The discovery of ever more specimens of feathered dinosaurs from China confirmed earlier suspicions, that the elongate arms of Maniraptora were indeed lined with large, complex feathers, equivalent to the primaries and secondaries seen in bird wings today. These dinosaurs could fly. But what is flight? It's important to realize that flight includes all kinds of ways to move through the air apart from leaping and plummeting to the ground. Today, there are many flying tetrapods, not just birds and bats, but also frogs with membranes between their toes, snakes that can spread their bodies out sideways, lizards with membranes down their sides and supported by elongate ribs, and a whole range of mammals with membranes extending between their arms and legs. These animals can all fly, but it is not the flapping flight of a bird or bat. However, they extend the distance they can travel when they leap from tree to tree, and that's a form of flight. It is commonly called parachuting if they are mainly slowing down their rate of descent, or gliding if they are mainly moving horizontally, but dropping as they go.

The feathered maniraptoran dinosaurs such as *Anchiornis* and *Microraptor*, then, could fly, in the sense of parachuting or gliding. There have been experiments on how they flew and how well, often using models in wind tunnels, and sometimes digital models. For example, Colin Palmer, an engineer who runs businesses selling sailed surfboards, carried out experiments on a model of *Microraptor*. He built a full-scale model from structural foam and coated it with resin to create a smooth finish, and lined the arms and legs with feathers plucked from modern birds, but arranged exactly as in the fossils. The model was tested in the 2 × 1.5-metre (7 × 5-foot) wind tunnel at Southampton University, normally used for testing the aerodynamics of cars and aircraft parts. Palmer mounted his model, and varied the wind speed, orientation of flow, and model posture, allowing the limbs either to dangle down or to stick out sideways. This did affect the glide, and the legs-down posture allowed the *Microraptor* model to glide further. In more detail, the best

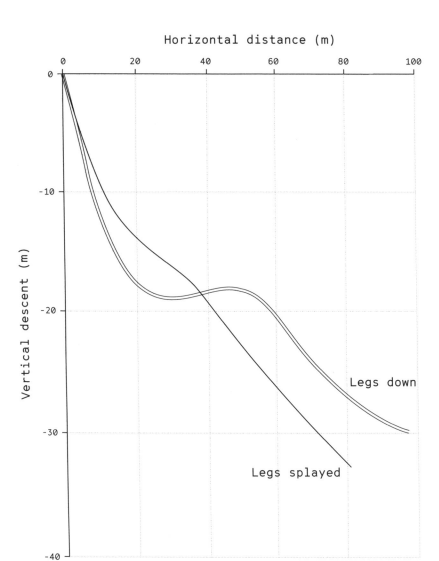

Horizontal distance (m)

Vertical descent (m)

Legs down

Legs splayed

Distances travelled by Colin Palmer's model
of *Microraptor* according to the position of
the legs – dangling down or tucked up.

policy for a gliding *Microraptor* was to jump off its perch with its legs
sprawled out to the sides, and then to let them drop down to the dangling
position as it entered the glide phase.

Why didn't *Microraptor* engage in powered flight? Palmer and
colleagues concluded that for moving around in the trees, at heights
of 20 or 30 metres (66 or 100 feet), there was no advantage in evolving
powered flight. With its feathered fore and hindlimbs and, one assumes,
decent eyesight and coordination so it could land efficiently without

crashing into trees, gliding was enough. The next step in flight, as seen in early birds such as *Archaeopteryx*, was to expend considerable energy in pumping the wings up and down, requiring massive pectoral muscles and high volumes of nutritious food. The Late Jurassic and Early Cretaceous woods were probably home to dozens of species of small theropods, sailing through the air, legs up or down, snatching insect prey from among the branches, or even on the wing, and escaping predators by their bold leaps.

Did bird flight originate from the ground up or the trees down?

In Chapter 4, we saw how feathers evolved, and gliding dinosaurs like *Microraptor* had all the specialized flight feathers seen in a modern bird. These Chinese fossils have settled a controversy about aerodynamics that raged when I was a student. Some experts declared that it was impossible for a glider to become a flapping flyer – they said the muscles and specialized bone joints were not there, and they pointed to modern gliders, such as lizards with extended ribs and skin, and flying foxes with skin flaps stretched between their arms and legs. Surely these gliders could not switch one aerodynamic structure for another, the arm-supported wing of a bird or bat? Those researchers also argued that birds evolved their flight from the ground up – leaping as they ran, and beating their wings to extend their leaps. I'd always thought this was a crazy idea.

Well, *Microraptor* and its relatives had proper wings, indeed four of them, lined with complex flight feathers. The step to *Archaeopteryx*, still accepted as the first bird, as we saw in Chapter 4, seems slight – increase the area of the wing a little so it can support the body weight, and strengthen the main muscle of the breast region, the pectoral muscle, so it can power the driving down-beat of the wing. *Archaeopteryx* might not have had the massive pectoral muscle of some modern birds, but its powers of flight were probably sufficient to lift it above the trees for short journeys of a kilometre or so. That would have been enough to escape from predators, or to hop to a new copse to exploit the crawling and flying insects in the tree tops.

Ground up or trees down, or something in between? I think the evidence of the Chinese fossils is hugely in support of the traditional 'trees down' model for the origin of flight. As any physicist will tell you,

'Use gravity, stupid!' Why would running dinosaurs fight against gravity by leaping up from the ground to snatch insects, and then somehow evolve ever more advanced abilities to fly? The maniraptorans most likely became small and extended their arms and powerful hands as direct adaptations for climbing and hanging on to trees. While their larger cousins, such as *Allosaurus* and ancestors of *T. rex*, were barging around on the ground in pursuit of large plant-eating dinosaurs, *Microraptor* and its kin were creeping around in the trees snatching insects and spiders, as well as lizards, frogs, and early mammals. The rich faunas of Jurassic and Cretaceous tree-dwellers have been revealed by recent collecting in the Burmese amber.

The 'trees down' theory for the origin of flight is that small, feathered theropods of the Jurassic pranced and displayed, using their bright feather colours to win mates, and maybe partly for camouflage in the dappled sunlight. Grasping the tree boughs with their powerful hands, they leapt from branch to branch, and those with slightly elongate feathers along their arms had an advantage by being able to take bold leaps across several metres of open space to another tree. Evolution favoured the jumpers, as they got more food or escaped more readily from predators. Longer feathers and longer, stiffer arms made the wing more effective and extended the leap. However efficient the gliding wing, though, gravity means the animal sinks towards the ground, so any gliding dinosaur that could pump its wing, even ever so slightly, could counter the downwards pull and extend its flight.

This might all be a bit simplistic, of course, and many biomechanics experts now focus on how flight performance might relate to growth, and how juvenile dinosaurs might have operated. For example, Ashley Heers and Ken Dial have explored how the flight performance of ground-dwelling birds today changes with age. In a series of experiments on the Chukar partridge, they found that very young specimens charge around and can use their undeveloped, short wings to aid them in jumping and taking runs at tree trunks.

These investigators have focused on a specialized mode of locomotion seen in these birds, called wing-assisted incline running, which is the best evidence to rescue some aspects of the old 'ground up' theory. Even juvenile Chukar partridges, which cannot yet fly, can enhance their speed and manoeuvrability by scampering and flapping their stumpy little wings at the same time. These authors then transfer the Chukar model to the early theropods, and make a case that these little dinosaurs charged around the Jurassic forests, running at trees to catch insects by propelling themselves, cat-like, halfway up the trunk, and then falling back. They argue that over time, this led to the evolution of full, flapping flight – and true birds.

Dinosaurs in films – do they get it right?

Back in 1995, I was invited by the BBC to act as one of six consultants for their new documentary series *Walking with Dinosaurs*. I was amazed to discover the close convergence between animators and physicists. In fact, this was a practical example of the overall shift of palaeobiology from speculation to science. If the film-makers had been working ten years earlier, we would have had little to bring to the table other than general observations of how large modern animals, such as elephants and ostriches, run and walk. Now we had biomechanics and digital models.

My PhD student, Don Henderson, now Curator of Dinosaurs at the Royal Tyrrell Museum in Alberta, Canada, was applying his background in pure physics to calculate the leg motions of dinosaurs from first principles. He decided to treat each element of the leg as an independent pendulum, hanging down from the hip, knee, and ankle joints. He then solved a set of three equations that described how far forwards and backwards the thigh, shin, and elongate ankle bones could swing, but constrained by the need to touch the ground neatly at the end of each stride. Most of his calculations ended with the foot below or above the ground, but that was clearly impossible. He accepted only those cases where the foot, correctly, touched the ground at the end of each step.

Henderson then went 3D, and visualized the walking hip with two legs from below. As the biped walked, the body rocked from side to side to keep balance, and the points of contact with the ground formed a triangle, from right to left to right. The secret was to keep the centre of mass within the advancing triangle. The centre of mass is the 3D pivot point that lay just in front of the hips in the lower belly region – balance a model on this pivot, and it would not tip forwards or backwards, or to left or right. If the centre of mass moved outside the moving stride triangle, the animal would fall over.

Having seen Don Henderson's stick-dinosaur animations, we went to the studios of Framestore, the animation company, located in the centre of London. In those days, the animation software they used to create live-action dinosaurs was enormously cumbersome, and required banks of the biggest Macintosh computers on the market to process the data.

The animated skin of *Allosaurus*, designed
to be slung over the stick model.

The animators used a four-step process to make the films, and they still
do. First, they make a storyboard and live film of the background scenes.
Where a dinosaur is to brush against a tree, a technician pushes the tree.
Where the dinosaur is to splash through a river, a helper dons great boots
and splashes through the water. The person is then edited out. Second,
the dinosaur motions are sketched as stick models moving through the
background. Third, grey cylinders are hung on the stick models like a
crazy suit of loose-fitting armour, to represent the flesh of the limbs and
body – they are hung in such a way that they stay put as the stick model
moves about. Finally, the skin is constructed like an old-fashioned lion
rug, with the skin opened out and painted in detail with all the scales,
feathers, and colours required. This 'rug' is then draped over the grey-
armour-suited dinosaurs, and the animators hope it sticks. In the early
days, the stick-dinosaur sometimes vacated his coat and ran off, leaving
it trailing a few frames behind.

That's not the point, though. The animators at Framestore had not
carried out the careful first-principles feasibility study that Don had, and
yet they got it right. As Mike Milne, the chief of the operation, explained,
'We've all spent time watching modern animals move, and so we know
what's plausible. The animal has to be light on its feet, balanced, and
moving in proportion to its intended body mass.' It's like the intuition

a ball player has in aiming or catching a ball – no need to calculate speeds and trajectories; you just know which way it's going.

When *Walking with Dinosaurs* was shown by the BBC in 1999, people were critical of all sorts of things, but nobody said the locomotion was implausible. This was to the credit of the animators, and their good sense in checking with biomechanics experts such as Don Henderson. Any deficits were down to money and computing – high-quality dinosaur animation costs a great deal, and the BBC did not have the budget of a Hollywood film producer.

The animators found that making the animals look heavy as they marched around the Jurassic landscapes was hard to achieve. Look at any film with animated dinosaurs – they often look as if their feet are hovering a few inches above the ground, not really thundering along and making deep footprints. The animators add deep, pounding sound effects and music, or puffs of dust, or they hide the feet behind a discreet curtain of plants. Still, the movements are astonishingly well rendered – and I haven't even mentioned how they get the sway of the body right. Sway right as you lift the left foot, left as you lift the right foot; they are making sure the animal obeys the centre of mass-stride triangle rule that Don Henderson had imposed, but they are doing that instinctively. The tail also sways in time with the undulating body, setting up rippling waves of balancing adjustments from front to tip of the tail. The head bobs from side to side and up and down as the animal walks – look at how a pigeon or pheasant nods its head as it walks. The muscles bulge in all the right places.

The impact of *Walking with Dinosaurs* can be appreciated from this publicity poster.

As for the quibbles and niggles by palaeontologists every time an animated dinosaur feature is produced – well, all I can say is, 'relax, they got it 99 per cent right, and they have mastered an astonishingly subtle set of commands to produce something plausible'. We would know for sure, and without a training in palaeontology or biomechanics, if they got it seriously wrong. Think of those old dinosaur movies with plasticene dinosaurs animated by stop-motion photography, or worse, the films with lizards sporting cardboard crests and spikes to make them look like some dystopian kind of yet-to-be-discovered dinosaur!

Pinning down how dinosaurs moved is important. We have seen how the earlier work by scientists and artists was compromised by changing views on their correct posture. A proper biomechanical approach shows that the dinosaurs must be correctly balanced, and that their bodies must have swayed from side to side and up and down as they walked. The fundamental point is that many earlier endeavours reconstructed impossible modes of locomotion, in which the animal would have pitched onto its nose or fallen sideways.

McNeill Alexander led the way, and Hutchinson took things into the modern world of computational digital modelling. These approaches have shown which postures and modes of walking and running are plausible, and which are not. The methods also allow researchers to calculate speeds and gaits – could a particular dinosaur run or gallop, or was its fastest speed achieved as a determined walk? Other questions about locomotion, such as the function of the arms of *T. rex*, and how dinosaurs could swim and fly, are moving forward, but much more remains to be discovered.

All of this brings palaeontologists directly in touch with film-makers, and mostly the film-makers get it right. The latest installment in the *Jurassic Park* franchise, *Jurassic World: Fallen Kingdom* (2018), continues the good work, with agile, heavy, plausible dinosaurs running across the screen – it's just a pity that the feathered ones are still shown without feathers, but that was a deliberate, stylistic decision by the producers.

John Hutchinson reflects on his recent work: 'We do not yet know whether the Mesozoic was played out in dinosaurian slow motion, or at modern speeds.' But he is convinced that what he does is testable science:

Indeed what we do in evolutionary biomechanics is science. We span from basic (but vital) descriptive research ('what is that?') to general questions ('how does that work?') to specific hypothesis-testing ('is that most likely right or wrong?'). And we emphasize that science is a continual process where we investigate our and others' past work with improved methods and evidence, where we can. We admit we're wrong when we must, and we admit uncertainty or variation where it exists, but also embrace subjectivity as a crutch that we must sometimes lean on to make progress. But the ground we walk on is that of science itself: clear, reproducible data and tools, a spirit of sharing and professionalism, and open-mindedness.

Chapter 9

Mass Extinction

Sixty-six million years ago, a great rock hit the Earth and wiped out the dinosaurs. The rock was an asteroid, essentially a small planet or a large meteorite. It measured up to 7 kilometres (4 miles) across, the size of Manhattan, and as it drove into the Earth's crust, just off the coast of the Yucatán peninsula in modern Mexico, it blasted out a deep hole and caused shattering of the crust to an even greater depth, and over a much wider radius, than the crater itself (see pl. xix).

The impact had a kinetic energy of more than 10 billion megatonnes. This is a thousand times the amount of energy contained in all the world's nuclear weapons arsenals. During the impact, the asteroid vaporized, sending powerful shockwaves downwards and sideways into the surrounding rocks.

After a second or two, once the asteroid had driven down as far as it would go, there was a massive reaction. Vast quantities of rocks shot upwards and sideways, creating a conical expanding crater. Larger blocks fell back into the crater and around its rim, but smaller boulders, melt materials, and rock dust formed from a mixture of the Earth's crust at the crater site and the asteroid itself rose as a huge plume and shot out sideways at high speed.

At the time, there would have been easterly winds around the equator, as today, caused by the westward rotation of the Earth, and these blew the plume of dust and rock fragments west of the crater. A blanket of rubble and bombs ejected from the crater formed outside the crater rim, and the dust was lofted into the upper parts of the atmosphere and travelled around the world.

The impact had two further effects. First, the melt rock from the impact site formed into small glassy beads, each about a millimetre across, and billions of these flew through the air to land in great mounds all over the landscape and in the sea. It was the speed of transit through the air that made them form into beads, as the molten rock cooled while twirling in the air.

The asteroid hit the Earth at the edge of the Caribbean Sea, and so it also produced great tsunamis, or tidal waves. The tsunamis formed walls

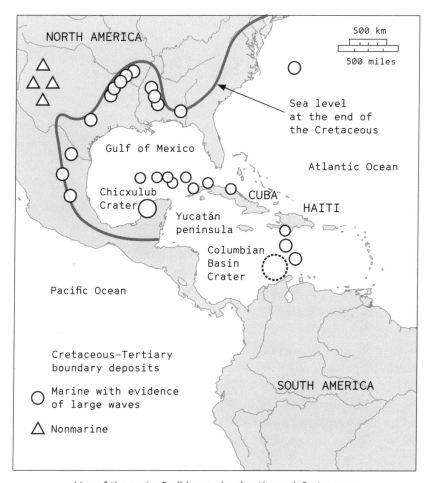

Map of the proto-Caribbean, showing the end-Cretaceous
shoreline, evidence of tsunami beds, and the impact site.

of water probably tens of metres high, and they travelled at the speed of
a jet plane, some 800 kilometres (500 miles) per hour. The Mexican and
Texan coasts, only a few hundred kilometres away, were devastated, with
the tidal waves ripping up rocks and life along the coast and dumping it
all in a chaotic pile. Further away, the tsunamis lost much of their height
and power, so dinosaurs in Europe would have seen little more than a
slight ripple on the beach.

The tsunami would have killed everything along the shores of
the proto-Caribbean, probably for several hundred kilometres inland.
Any dinosaurs dipping their toes in the water would have been rudely
awakened, flung in the air, and hurled to the ground. Other killing
effects of the impact included the rain of huge rocks from the sky –

but that probably didn't kill too many dinosaurs, since most of these rocks would have fallen back into the sea. Yet the smaller gravel-sized particles, including the glass melt beads, did travel over land and they would have peppered the landscape, and any dinosaurs in their path, with scattershot punches as if from a huge shotgun.

Then came the second pulse, beginning a few seconds after the debris cone. A huge fireball shot upwards from the crater site, composed of vaporized material from the asteroid and taking heat from the huge energy of the impact. The fireball expanded sideways, after the debris of the first phase had settled. It set fire to all plants and animals in its path, leaving a blackened landscape. Again, like the physical blast, the fireball could not have encircled the Earth, but it devastated North America and the Caribbean. So, the fireball was a fourth potential killer (after the rocks, melt beads, and tsunami): bad news if you were in its way, but still not enough to cause worldwide extinction.

It was the dust, drifting seemingly innocently in the upper reaches of the atmosphere, many kilometres up, that was the real killer. As the tsunami, rockfalls, and wildfires were sweeping out from the crater, the huge black dust cloud blew rather passively with the winds around the northern hemisphere. It probably covered some of the southern hemisphere, maybe all of it. Global wind patterns do not guarantee that the whole globe would have been shrouded.

It may have taken a few days after the impact for the dust cloud to reach its full extent. It probably showered fine dust particles on the Earth all the time, but would have taken years to dissipate completely. This was no innocent cloud, though. It contained millions of tonnes of dust. As it spread and thickened, the Earth beneath was thrown into total blackness. With the sun's rays entirely blocked, no light or heat could get through for perhaps a year. Now that might be the real killer.

In addition to all these details, we know that the impact happened in June...more about that later.

These cataclysmic events 66 million years ago set the shape of the modern world, including the dominance of modern ecosystems by birds and mammals. And yet the whole scenario was unknown when I learned geology in the 1970s – indeed, if anyone had talked about such a catastrophe they would have been ridiculed. We have thus seen an astonishing switch in our understanding of the mass extinction that finished the dinosaurs, from rejection and speculation to a substantial body of established scientific knowledge. How did this happen?

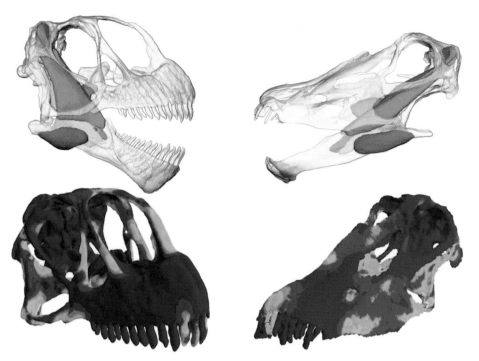

IX The skulls of *Camarasaurus* (left) and *Diplodocus* (right) showing reconstructions of the jaw muscles (above) and the application of different loads (below). In the loading illustrations, warm colours (red, yellow) indicate high stress and strain.

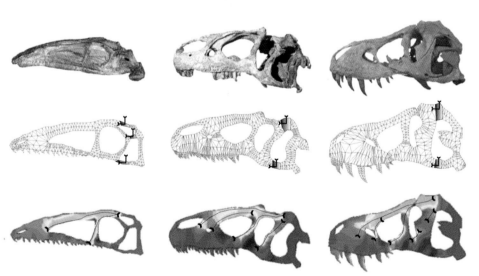

X Comparison of the skulls and their engineering properties of *Coelophysis* (left), *Allosaurus* (middle) and *Tyrannosaurus* (right), showing photographs of the skulls (top row), the surface mesh models (middle row), and the loaded models (bottom row). Red shows high stress and strain, green low values, and arrows the ways in which forces are distributed.

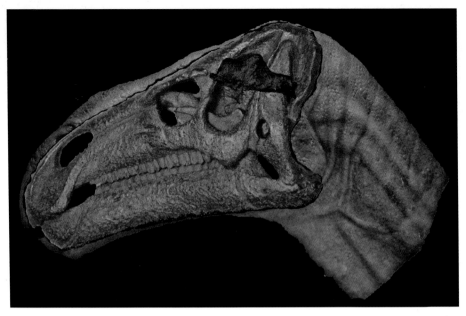

XI Cutaway of the skull of *Iguanodon* with the brain imaged in position.

XII The skull (above), brain (below) and semicircular canals (in pink) that form the middle ear of *T. rex*.

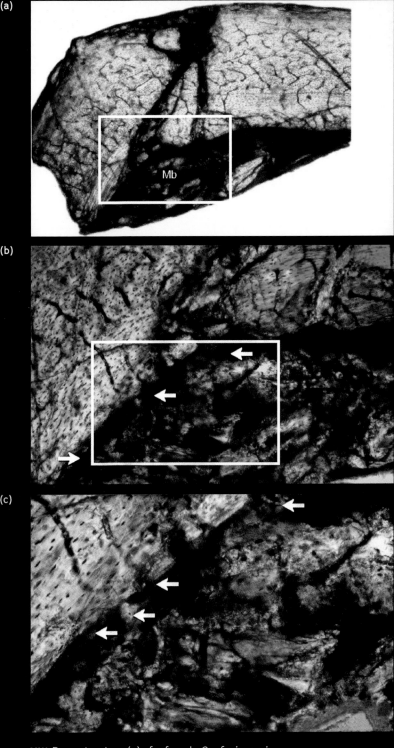

(a)

Mb

(b)

(c)

XIII Bone structure (a) of a female *Confuciusornis*, showing medullary bone (white arrows in b and c).

XIV A mother *Maiasaura* (meaning 'good mother lizard') looks dotingly at a nest of her babies in the Wyoming Dinosaur Centre, Thermopolis.

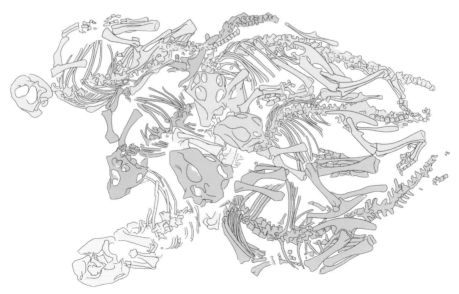

XV Juvenile dinosaurs used to hang out together. A block containing six juvenile *Psittacosaurus*. Bone histology and growth rings show they are all two years old, except number 1 (in pink) which is three.

XVI Side view of the skull of a juvenile *Massospondylus*, based on CT scans.

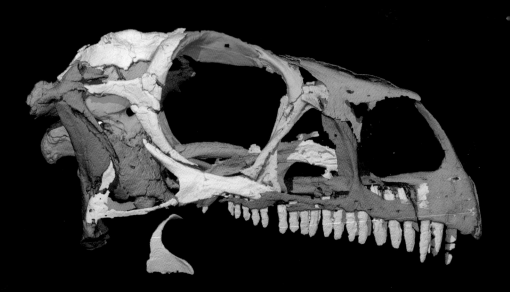

XVII Side view of the skull of an adult *Massospondylus*, based on CT scans.

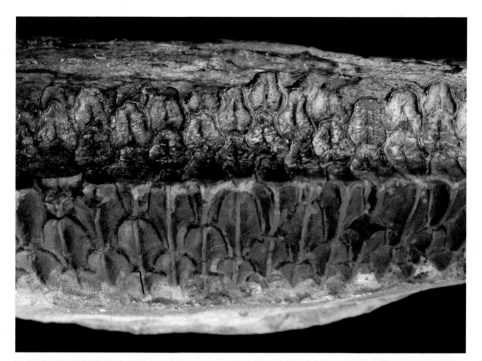

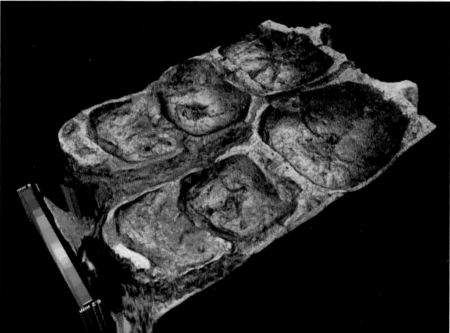

XVIII The unique teeth of hadrosaurs. Replacement teeth line up below the set in use (above). The dental tissues (below) show multiple folds, as in a bison or horse's teeth.

XIX The asteroid impact that ended it all, 66 million years ago. Artist's impression of the huge rock from space as it hit the proto-Caribbean Sea over what is now southern Mexico (above), and the resulting double-ringed crater (below).

The road to accepting mass extinction

This whole dramatic story of devastation now seems clear and it is supported by a great deal of evidence. When I was a student, though, mass extinction was not considered. We heard that dinosaurs had died out gradually, over millions of years, and so, in some way, the transition from the Cretaceous period to the Palaeogene period 66 million years ago was thought to be gradual. Looking back, this seems amazing – how could geologists and palaeontologists have misunderstood what the rock and fossil records showed?

I believe there were three reasons, back in the 1970s, why geologists and palaeontologists kept well clear of mass extinctions: fear of catastrophe, fear of numbers, and fear of ridicule.

Geologists had been taught by Charles Lyell in the 1830s that catastrophes did not occur. In his seminal *Principles of Geology*, he set out with lawyerly skill, and based on extensive field work, the new geological science. The science was what he had seen in his native Scotland, in England, France, and Italy, and his main argument was that geologists must use pure observation and a view of modern processes to interpret the history of the Earth. To add piquancy, he used his training as an advocate to identify an opposing view, termed catastrophism, which he then exaggerated in order to bolster his principle of uniformitarianism: that geological processes occurred at a uniform rate, as we can observe today. Lyell was the supreme rationalist, and he painted his opponents, including Georges Cuvier and others, as dangerous, wild, relying on supernatural explanations – in the case of Cuvier also French, and conveniently deceased as the last volume of *Principles* went to the press in 1832.

As students, we were taught Lyell's uniformitarianism, as are all geologists today. It is obvious that we should observe how modern volcanoes, rivers, and beaches work, so we can interpret the ancient rocks laid down by these agencies. Lyell went further, though, and claimed that not only were the processes the same, but so too were the magnitudes or scales of such processes. He argued that volcanoes in the past were no greater than today – but critics now point out that this is perverse. Why? Relying solely on human experience is an unnatural narrowing of the frame of reference – and we now know that there were many huge volcanic eruptions and meteorite impacts in the past, far greater in magnitude than any human has observed, or at least observed and written about. Nonetheless, Lyell's views of uniformitarianism held a tight grip over geology until the 1980s.

Palaeontologists have always been afraid of numbers. I remember, as a young lecturer, attending a special meeting of the Royal Society in London in 1988, where the theme was extinction, and we heard talks by twenty world experts, some flown in from the United States. The most notable invitee was David Raup from the University of Chicago, and he gave what I thought was a perfectly reasonable talk about the evidence for extraterrestrial causes of extinction. He was looking at how to interpret species extinctions in the fossil record. He deployed a numerical approach, one that used repeated random sampling of simulated data to show how missing rock layers and missing fossils could give misleading results. At the end of his talk, a British professor stood up and said, 'We don't want these kinds of crazy ideas brought in from North America', and more of the same. I was astounded, and should have defended Raup, but the attack was received with some humorous and perhaps supportive snorts from the audience. Raup, an extremely brilliant and gentle man, who had led the field for decades in showing smart ways to turn palaeontology into science, swore never again to visit Britain.

What was the logic behind this mindless attack? We had a bit of nationalism ('we don't want foreign ideas here'), a bit of protectionism ('these are my fossils and I'm the expert on them; you can't use my data'), and of course fear of numbers. It's clear, though, that Raup was right and his critic was wrong. Palaeontology ought to be as much of a science as any other, and mysticism or unfounded claims of authority have no place.

The third fear was fear of ridicule, which relates to catastrophism, but also to the long history of theorizing about the death of dinosaurs. I have counted more than 100 'theories' for dinosaur extinction that have appeared in scientific journals since the 1920s (see Appendix). These ranged from environmental catastrophes (it got too hot or too cold, too wet or too dry), to dietary issues (caterpillars ate all the plants, mammals ate all the dinosaur eggs, or new plants gave dinosaurs constipation), to mystical assumptions (dinosaurs were too big, they got arthritis, their brains shrank, their horns and headshields were too unwieldy, they were too weird to evolve, they got AIDS). Impact by meteorites or comets seemed just as daft, so it was thought to be safer to keep your head down and say that the study of extinctions is too dangerous for any sane researcher.

The concept of mass extinctions is now much better understood by scientists and the public. In fact, mass extinctions can be claimed as one of the most important discoveries in the Earth sciences. Geologists and palaeontologists have unique access to the body of data about them, and mass extinctions cannot be predicted from any study of modern organisms. They were events of profound scale and importance in evolution. When one thinks about it, they have a positive side too – we always attribute the success of 'modern' groups, such as birds and mammals, including ourselves, to the events that saw the end of the dinosaurs, and freed the world for major ecological restructuring.

The change in viewpoint, from a fear of talking about mass extinctions to acceptance, happened about 1980, and it took a Nobel-prize-winning physicist to shake the palaeontologists out of their cocoon of complacency.

The impact of the impact theory in 1980

The hammer blow fell on 6 June 1980, eight years before the London meeting at which Raup was so snidely dismissed. I was then a doctoral student in Newcastle, reading as widely as I could about dinosaur evolution and extinction. On that day, a paper entitled 'Extraterrestrial cause for the Cretaceous–Tertiary extinction' was published. It said:

> A hypothesis is suggested which accounts for the extinctions and the iridium observations. Impact of a large earth-crossing asteroid would inject about 60 times the object's mass into the atmosphere as pulverized rock; a fraction of this dust would stay in the stratosphere for several years and be distributed worldwide. The resulting darkness would suppress photosynthesis, and the expected biological consequences match quite closely the extinctions observed in the paleontological record.

The paper was led by Luis Alvarez, who had won the Nobel prize in 1968 for his invention of a means to image interactions between particles in the newly developed hydrogen bubble chamber. He was well respected for his brilliance and skill at building impossible pieces of equipment in the laboratory. He also had a brusque approach to scientists when he thought they were poor thinkers. For example, in a telephone interview reported by a *New York Times* writer in 1988, he said: 'I don't like to say bad things about paleontologists, but they're really not very good scientists. They're more like stamp collectors.' Understandably, remarks such as these, whether true or false, did not endear him to the dinosaur community.

The Alvarez team also included his son, geologist Walter Alvarez, as well as geochemists Frank Asaro and Helen Michel. The discovery hinged on Luis Alvarez's invention of a means to measure vanishingly tiny amounts of the element iridium, chemically related to platinum. Iridium occurs in minute quantities in soil and rocks, and it can be a small component of some volcanic lavas, but is much more abundant in space, by a factor of 720 times. Therefore, most of the tiny quantities of iridium found on the Earth's surface comes from extraterrestrial sources, primarily the continuing shower of small meteorites (tektites) over the surface of the Earth.

The idea of Alvarez *père et fils* was to use the steady (but minute) rain of iridium as a chronometer against which rocks could be dated. For a long time, geologists had realized that rock thickness does not equal time, for two reasons. First, some rocks are deposited very fast and some very slowly – an extreme example comes from the deep ocean, where muds may typically accumulate at rates of only a few centimetres per century, but can be interrupted by catastrophic turbidity flows, sometimes triggered by earthquakes, in which hundreds of metres of sand and rock may be dumped in a single day. Second, we have the gaps between rock layers, and we have no idea how long a gap might be. If there were

Above: Luis (left) and Walter Alvarez with his hand on the Cretaceous–Palaeogene boundary in the Gubbio section, Italy. *Opposite*: The iridium spike.

a yardstick of time, such as a measurable influx of iridium dust, then geologists could at least tackle the first issue.

Walter Alvarez chose a rock section near Gubbio in central Italy, near Perugia, where he knew there were hundreds of metres of marine limestones of latest Cretaceous and earliest Palaeogene age, which were well dated by microfossils. The sampling showed similar levels of iridium near the bottom and top of the section, suggesting in fact that the limestones had been deposited in a steady manner. But in the middle, at the Cretaceous–Palaeogene boundary, now dated at 66 million years ago, there was a spike where values shot up to ten times normal – that is from 0.6 parts per billion to 6 parts per billion (these tiny quantities show the need for a sensitive measuring instrument). Now, this is where the first smart thing happened – the Alvarezes took a sideways leap. If they had stuck to their working hypothesis, they would have said this boundary layer was highly condensed, meaning that it took ten times longer to

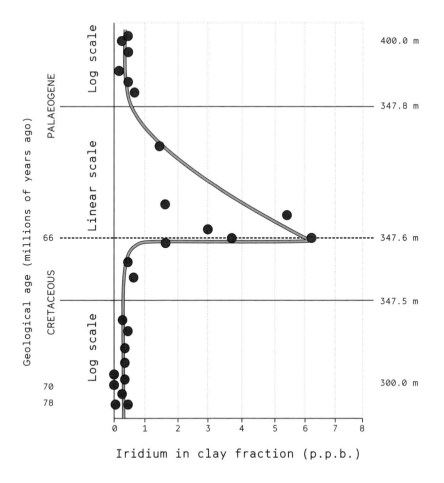

deposit that particular 1 centimetre (⅜ inch) of sediment than above or below, and so the iridium level shot up ten times. Instead, they daringly said that this indicated the sudden and rapid arrival of a huge amount of iridium from outer space: thus a vast meteorite.

In their paper, Luis Alvarez and colleagues then used this one observation as a basis to erect their hypothesis. (They did cross-check it against one other section, at Stevns Klint in Denmark.) The reasoning was that if a large meteorite, an asteroid, killed the dinosaurs, it must have thrown up a sufficiently large cloud of dust to encircle the Earth, and they did a back-calculation from that assumption. Here is their formula:

$$M = 0.22f/sA$$

where 'M' is the mass of the asteroid, to be worked out from the other factors, which are known: 's' is the surface density of iridium just after impact (8×10^{-9} grams per square centimetre); 'A' is the surface area of the Earth; 'f' is the fractional abundance of iridium in meteorites (0.5×10^{-6}, known from modern meteorites); and 0.22 is the proportion of material from the 1883 eruption of Krakatoa that entered the stratosphere. So, they calculated M = 34 billion tonnes, equivalent to an asteroid diameter of 7 kilometres (remarkably, precisely the size estimated when the actual impact crater was eventually found), and predicting a crater twenty times as large, say 150 kilometres (over 90 miles) across.

This, then, is the origin of the model that a 7-kilometre-wide (4-mile) asteroid hit the Earth, vaporized, and threw an immense cloud of ash into the upper atmosphere, which encircled the globe and blocked out sunlight, so photosynthesis in green plants ceased, and life consequently died on land and in the oceans.

The publication caused uproar. Geologists were incensed – who was this crazy physicist telling us what to think? We all know asteroids never hit the Earth; this goes against Lyell and uniformitarianism. Palaeontologists, with their hangups, were defiant too. Bob Bakker said what many were thinking: 'The arrogance of those people is simply unbelievable. They know next to nothing about how real animals evolve, live, and become extinct. But despite their ignorance, the geochemists feel that all you have to do is crank up some fancy machine and you've revolutionized science.' Well, Bakker was wrong, as were many (perhaps most) other palaeontologists and geologists at the time. The impact really did happen, as we now know, based on hard evidence from field research, as we shall see.

Periodicity and nuclear winter

The idea of an asteroid impact 66 million years ago immediately spawned another, perhaps more startling consequence. If an impact happened once, why not many times? The suggestion was made in 1984 by David Raup and his colleague Jack Sepkoski, based on their preliminary analysis of the fossil record. They focused on the past 250 million years, and plotted a measure of extinction through time. They expected to see times of low and high extinction, but were surprised to see what looked like a regularly repeating signal of extinction peaks. The raw measurements suggested that the peaks repeated every 26 million years, but Raup and Sepkoski had to test this, so they applied a numerical analysis to assess whether the pattern could have arisen by chance or not – the repeat pattern, termed periodicity, was significant to a high degree of probability.

Astronomers were excited by the Raup and Sepkoski data because any phenomenon with a periodicity, or repeated pattern, on the scale of 26 million years was likely to have an astronomical driver. Astronomers considered three main theories: that the entire solar system was tilting up and down like a wobbly plate; or that there was a sister star to the sun called Nemesis; or that there was a tenth planet, called Planet X, lying at the edges of the solar system. In all three cases, the perturbations affected the outer fringes of the solar system, where the Oort cloud of comets is

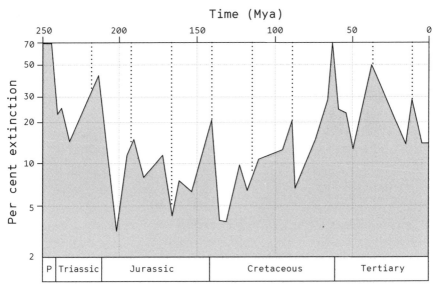

The periodicity in mass extinctions of 26 million years,
as proposed by Raup and Sepkoski.

located. The perturbation sent comets flying into the heart of the solar system, and one or more would have hit the Earth.

If the periodic pattern were true, the next flurry of impacts could then be predicted. The last one happened 14 million years ago, so the next one would be in 12 million years' time – a great way to test the hypothesis. I remember, with fellow palaeontologists, being amazed, and slightly awed, by the fact that a relatively straightforward synoptic diagram taken from our knowledge of the fossil record could have set loose such amazing speculations about the functioning of the Earth and the universe.

From the start, geologists and palaeontologists pointed out flaws in the line of reasoning. The periodic signal depended on a particular dating of the geological record, and slight revisions would break up the tightness of the 26-million-year period. They also noted that the last event is barely supported by any data – and yet that should have been the clearest to see in the rock record, because it is nearest to the present day. Further, the matching of events to the regular periodicity broke down in the Jurassic and Cretaceous. The debate rumbles on, with revivals of the idea in papers in 2016 and 2017, but most have abandoned the idea of periodicity.

One other consequence of the Alvarez model has had more traction, and that is the idea of nuclear winter. Three years after publication of the Alvarez paper, several climatologists began to speculate about the effects of all-out nuclear war. Richard P. Turco coined the term 'nuclear winter' in 1983 to describe the main outcome of mass bombing, which would be the lofting of ash into the upper atmosphere that would blot out the sun, leading to freezing conditions as the warming effect of sunlight was removed. Quickly the climatologists, modellers, and futurists saw the parallels with the Alvarez extinction model, and the assumptions are all now accepted, both for the impact at the end of the Cretaceous, and for the consequences of a similarly massive energy release from explosion of the Earth's nuclear arsenals. Periodicity may have bitten the dust, but nuclear winter and impact killing of dinosaurs survived scrutiny. Then the crater was found.

The killer crater

The impact theory, periodicity, and nuclear winter idea set scientists, and the public, talking, and in 1985 the BBC made a Horizon programme about the proposed end-Cretaceous asteroid impact. The journalists asked the rather obvious question: where was the crater, the smoking gun? At the time, the geologists could say little more than that the crater might well have been lost somehow – which wasn't a hugely satisfactory answer. Even then, however, the trail of detective work was pointing to where the crater must be.

Geologists had noticed that there were strangely perturbed rock units at the Cretaceous–Palaeogene boundary in rock sections throughout coastal areas in Mexico and along the Brazos River in Texas. In the midst of orderly, flat-lying beds of limestone and mudstone were levels where the limestones had apparently been ripped up and dumped higgledy-piggledy. These were called storm beds, or even tsunami beds, or tsunamites. The idea was that the rocks had been torn asunder and dumped along the coastline of the proto-Caribbean, which lay inland along a line arcing through Mexico and the southern United States. If these geologists were right, then it implied there had been an impact out in the ocean, and a huge shock wave radiating outwards, in the form of a tsunami front, many tens of metres high, that beat on the shores and smashed up the freshly deposited rock layers.

The next clue came in 1991 from close study of a rock section across the Cretaceous–Palaeogene boundary at Beloc on the Caribbean island of Haiti by geologists Florentin Maurrasse and Gautam Sen. They noted that the boundary bed was 72.5 centimetres (28½ inches) thick, not 1 centimetre, as at Gubbio and Stevns Klint in Europe. Geologists read this great thickness to indicate that the source of the asteroid impact was not so far away. In the lower layers, the sediment was stuffed full

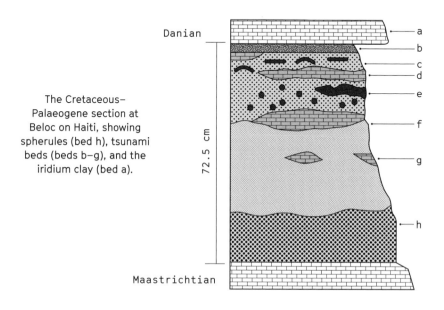

The Cretaceous–Palaeogene section at Beloc on Haiti, showing spherules (bed h), tsunami beds (beds b–g), and the iridium clay (bed a).

of glassy spherules of exotic geochemical composition – they were interpreted as impact glasses, which were formed by high-pressure and high-temperature effects at an impact site some distance away. These small glass beads were thrown high in the air, and carried in the atmosphere, together with other impact debris, for some 1,000 kilometres (620 miles). Glass spherules are commonly thrown out by volcanoes, but they have the chemistry of an igneous rock, such as basalt or andesite, matching the molten lava. The Beloc glass beads, bizarrely, had the chemistry of limestone and natural rock salt – in other words, they had come from the melting of such rocks, and they provided a direct clue to the nature of the then-unknown impact site.

Higher in the Beloc section, the researchers noted a layer of tsunamite, with perturbed limestone rocks thrown up, and finally the uppermost, 1-centimetre-thick, dust layer, with enriched iridium. This top layer is called the impact layer, and this is all that is found further from source, such as at Gubbio. The researchers identified that the Beloc boundary beds showed two thick layers at the base that could only have been generated in a location close to the impact, and they interpreted these as showing two separate phenomena, multiple glass beads hurled through the air and falling rapidly into the sea or onto land, and the tsunami beds, later. This indicates that one impact occurred and it generated two shock waves, one that rushed through the air first with the glass beads, and the second moving more slowly through the water.

In fact, the geologist Alan Hildebrand and colleagues had already identified the actual crater and they published it a few months later, in 1991. Hildebrand had located it in old borehole records made in the 1960s by Pemex, a Mexican oil company, drilling into an anomalous structure they detected deep beneath the Yucatán peninsula near the village of Chicxulub. Pemex quickly discovered this was not an oil trap when they hit melt rocks, and so they abandoned their efforts. Hildebrand, however, was looking for the crater, and he identified the melt rocks as typical of high-pressure impacts, where the meteorite hits the Earth, and smashes deep into the crust, vaporizing as it goes, melting the bedrock.

Hildebrand's initial geophysical survey was confirmed by later work. Samples of meltrock gave an exact age matching the Cretaceous–Palaeogene boundary, and further geophysical survey and drilling in 1997, 2002, and 2016 have given the detail. The crater is indeed formed in latest Cretaceous limestones and rock salt, as predicted from the Beloc glassy spherules. The seismic profiles show that the centre of the crater comprises an inner ring of shattered rock with a diameter

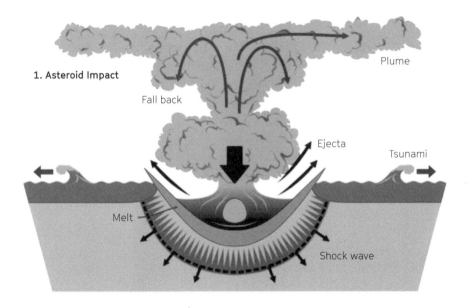

1. Asteroid Impact

Plume

Fall back

Ejecta

Tsunami

Melt

Shock wave

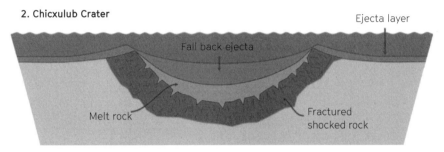

2. Chicxulub Crater

Ejecta layer

Fall back ejecta

Melt rock

Fractured
shocked rock

The sequence of events when the asteroid hit the Earth,
first penetrating and vaporizing (1), and then bouncing
back, leaving a crater and debris fields (2).

of about 80 kilometres (50 miles). This is a so-called 'peak ring', an
inner mountainous ring seen in large craters on other planets. Round
the edges of this crater is a zone of slumped rock, the terraced zone, that
extends to a diameter of 130 kilometres (81 miles), marking the limit of
collapse of the walls of the original crater. Further out still, at a diameter
of 195 kilometres (120 miles), the investigators found a major slope that
extends over 35 kilometres (21 miles) deep into the Earth's crust and into
the underlying mantle of the Earth. This outer ring is like features seen
in craters on other planets, such as Venus, and it's the first time such a
feature has been identified in a crater on Earth.

The sequence of events documented by the Chicxulub crater, and
predicted by Alvarez and colleagues in 1980, is that the 7-kilometre
asteroid smashed into the Earth's crust, burrowed deep, and vaporized.

The classic droplet rebound, showing how the rebound from an impact can lead to a similar central structure that then collapses.

Within seconds, the equal and opposite recoil happened, sending vast energy vertically upwards, and radiating outwards to form the outer crater wall. The recoil headed upwards, bringing the edges of the crater hundreds of metres into the air. Because of the size of the crater and the height of its walls, it collapsed rapidly under gravity, forming the outer and inner walls. The peak ring in the middle is part of the recoil. Just as the rebound of a water droplet hitting the surface of water sends a little spout of water upwards, which then falls back, so too with the rebound phase and formation of a crater.

The physics of the crater and the physical consequences, in terms of the blocking out of the sun, and consequent ending of photosynthesis, and global freezing, are pretty clear. How life was killed is less certain: but these effects would be enough to finish off the dinosaurs. In addition, there were huge volcanic eruptions in India, the so-called Deccan Traps, which began half a million years before the asteroid impact, and these eruptions would have driven regional-scale warming and acid rain. Also, climate cooling had begun about 30 million years before the impact, and this would have put pressure on the survival ability of dinosaurs, which probably preferred warm-climate conditions. The interplay of these longer- and short-term crises before the impact is still debated.

But, as noted at the start of this chapter, there is no debate about when the asteroid hit – even down to the actual month.

How do we know the impact happened in June?

Jack Wolfe, a veteran palaeobotanist working for the United States Geological Survey, had studied Late Cretaceous fossil plants all his career. He was looking at a rock section in Wyoming at a site called Teapot Dome in the 1980s, when he realized he was seeing the exact minute-by-minute story of the whole end-Cretaceous impact. The Teapot Dome section spanned the boundary; there was the boundary clay, only 2 centimetres thick or so, but Wolfe found he could pull it apart millimetre by millimetre to see what had happened. And, with his unrivalled knowledge of fossil plants, he could also pin down the temperatures throughout the event.

The Teapot Dome section records events in an ancient lily pond. In the latest Cretaceous, the lilies seemed to be flourishing, and Wolfe found dozens of leaves and stems that were buried as the pond silted up. At the Cretaceous–Palaeogene boundary, he found first a thin layer with glass spherules and dust. This marked the arrival of the first phase after the impact blast. After a few millimetres of this, there was a layer with dead lily leaves. Under the microscope, Wolfe saw that the cells in the leaves had burst. This was unequivocal proof of freezing – the sap had been shock-frozen. Ice occupies more space than water, so the ice crystals had pierced the cell walls.

Above the freezing layer was a second dust layer, and this one contained the iridium spike. Then sedimentation returned to normal, the lilies recovered, and temperatures returned to 25 degrees or so. Palaeobotanists can measure ancient temperatures quite accurately if the fossil plants have modern relatives, because plants have quite specific temperature and water requirements, and these can be assumed for their ancient relatives. It is a very accurate way of determining Cretaceous climates. Still, how did Wolfe know the impact happened in June?

He used Lyell's principle of uniformitarianism, comparing the fossil case with the modern world. The fossil lilies at Teapot Dome happen to be close relatives of the modern pond lily *Nuphar* of the family Nymphaeaceae. The lilies were instantly frozen at a particular stage of their development, and from the state of their buds and flowers, and by comparison with modern lilies of the same genus, Wolfe saw that the freezing had happened in June. It's a neat example of impressive detective work by a scientist using the principle of uniformitarianism.

Was the death of the dinosaurs sudden or gradual?

We now know that all the dinosaurs – except for some bird species –
died out after the cataclysm of 66 million years ago. Did they vanish
with a bang or a whimper? In other words, were they in fine fettle before
the asteroid struck and wiped them out, or are there signs they were in
decline anyway? In support of the sudden-disappearance model are well-
studied rock units that go right to the end of the Cretaceous, such as the
famous Hell Creek Formation of Montana, where dinosaurs including

Genus: **Triceratops**

Species: *horridus*

Tyrannosaurus, **Triceratops**, and **Ankylosaurus** are found right up to the very end of the Cretaceous (see pl. iii). There's no sign that individual faunas were becoming less diverse, that species were dropping out. Is it, however, simply a choice between a long decline or instant disappearance, or could there be a third model?

Our evidence is that it was a bit of both. We found a first clue in a study in 2008, led by my then student Graeme Lloyd, and introduced in Chapter 2. Our aim was to build a 'supertree' of dinosaurs, a fancy term for a phylogenetic tree of all dinosaur species that summarizes the best of

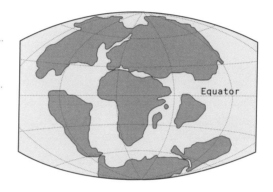

Named by:	Othniel Marsh, 1889
Age:	Late Cretaceous, 68–66 million years ago
Fossil location:	United States, Canada
Classification:	Dinosauria: Ornithischia: Ceratopsia
Length:	8 m (26 ft)
Weight:	14 tonnes (30,865 lbs)
Little-known fact:	*Triceratops* is the official state fossil of South Dakota, and the official state dinosaur of Wyoming.

Genus:	*Ankylosaurus*
Species:	*magniventris*

current knowledge. Graeme crunched data from 200 published trees, and we produced a supertree of some 420 species. We then calculated rates of diversification, and found to our surprise that dinosaurs had done most of their fast evolving in the first 60 million years of their history. During the Late Jurassic and Cretaceous, nothing out of the ordinary happened, and we argued that dinosaurs had more or less run out of steam in their last 50 million years, except for two specialized plant-eating groups, the duck-billed hadrosaurs and the horn-faced ceratopsians.

We picked up this idea again in a further investigation in which I was involved, in 2016. We wanted to find out the deep-seated evolutionary

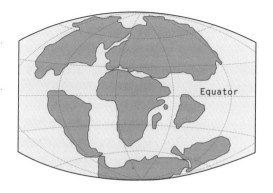

Equator

Named by:	**Barnum Brown, 1908**
Age:	**Late Cretaceous, 68–66 million years ago**
Fossil location:	**United States, Canada**
Classification:	**Dinosauria: Ornithischia: Thyreophora: Ankylosauridae**
Length:	**7 m (23 ft)**
Weight:	**4.8 tonnes (10,584 lbs)**
Little-known fact:	**The tail club weighed about 20 kg (44 lbs) and the impact force could be as much as 2,000 newtons, equivalent to a mass of 200 kg (440 lbs).**

dynamics of dinosaurs throughout their history. With colleagues Manabu Sakamoto and Chris Venditti from the University of Reading, we put together an even larger supertree of all dinosaur species, and dated it as accurately as we could. We then ran calculations to work out whether speciation and extinction rates were stable, rising, or falling through the Mesozoic. We were looking for one of three possible outcomes: that overall the balance of speciation and extinction gave ever-rising values, or levelling off, or declining values.

We used Bayesian statistical methods, which involve seeding the calculations with a starting model, and then running the data millions

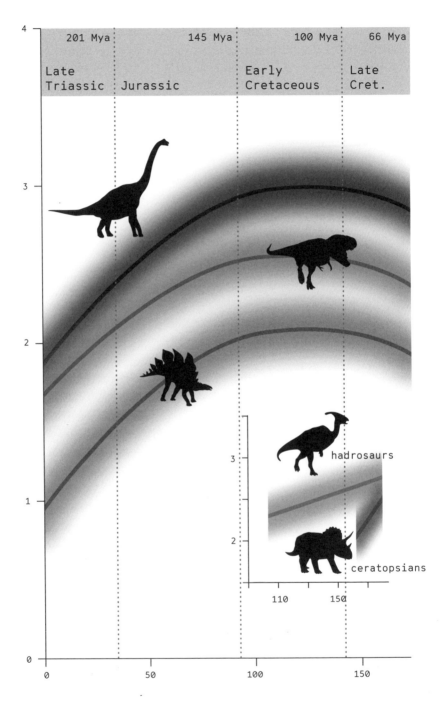

All three groups of dinosaurs – sauropodomorphs (top), theropods (middle), and ornithischians (bottom) – show a long-term decline in the last 40 million years of the Cretaceous.

or billions of times to assess how well the starting model fits the data, allowing for every possible source of uncertainty, and repeatedly adjusting the model to make it fit better. In this case, Manabu modelled uncertainty about dating the rocks, gaps in the record, accuracy of the phylogenetic tree, and many other issues. The results were unequivocal: for all dinosaurs, and for each of the three major dinosaur groups (theropods, sauropodomorphs, ornithischians), they showed buoyant evolution until 100 million years ago, with speciations exceeding extinctions, and then they all tailed off about 40 million years before the end, turning down into decline. Only two groups, as we had found before, the duck-billed hadrosaurs and the horn-faced ceratopsians, showed increasing diversity.

Our 2016 paper has been controversial – some researchers have got it, but others have misunderstood. Our result does not say that there are not individual locations, such as the Hell Creek Formation, where diversity remained locally high, nor that all groups of dinosaurs were in free-fall. There's a time-scaling aspect as well – we see a gradual decline over 40 million years, but at regional scales and over short time spans, diversity would continue to rise and fall, depending on local conditions.

There is one broader implication, and that concerns whether dinosaurs could have continued evolving. It was a common pursuit once to indulge in such imaginative 'what if' thinking: if the asteroid had not hit, what would the world look like today? One Canadian palaeontologist even reconstructed the modern dino-man, a kind of humanoid dinosaur with a big brain, no tail, and human-like posture. However, our work suggests that, even if the asteroid had not hit, the dinosaurs would probably not have survived to the present day. Cooling climates, changing vegetation, and changing configurations of the continents might have done for them anyway by 50 or 40 million years ago.

The killing worldwide

The sequence of events following the asteroid impact was outlined at the start of this chapter. We do not yet know exactly how quickly the crisis spread over the world, but the effects differed from place to place. Within a few hundred kilometres of the strike, the impact would have stimulated an immediate blinding flash of light, followed a few seconds later by a massive earthquake, as the asteroid drove into the crust and vaporized. Then the debris cone shot sideways and in an ever-expanding circle. Within 500 kilometres (310 miles) of the impact, the power of the shock wave would have killed everything, including any dinosaurs, such as *T. rex*, *Triceratops*, or *Ankylosaurus*. The rocks carried by the blast would also have pummelled the trees and animals to the ground. Further away, the larger rocks would have already fallen to the ground, but the expanding shock wave continued for thousands of kilometres. Around the crater, the oceans mounted into a huge tsunami that probably did not kill life in the sea, except when it came near the shore. There, the hungry wave sucked water back from the shallows, exposing coral reefs and flapping fishes and marine reptiles, and then crashed onto the shores of the proto-Caribbean, along a broad arc through eastern Mexico, and swinging up through Texas and round the southern United States to Florida. Any animals on the shore, including dinosaurs, would have been drowned and their carcasses thrown into piles with loose rocks and trees.

The shock wave and following fireball probably shot across most of North America, killing plants and animals as they went. Further afield, for example in Europe, Asia, and Africa, the flash of the impact might have been felt, and then, some hours later, the sky would have changed colour as the dust spread outwards, high in the atmosphere. Perhaps the dinosaurs were not much affected during the first day or two, but then the build-up of dust would have made day into night, and the darkness continued perhaps for weeks on end.

Without light, the dinosaurs remained in a torpid state, lolling on the ground, dozing as they always had during the night. But this time, the warmth of the rising sun did not come, and they slept on, most of them perhaps passing without knowing to death. The lack of light killed most animals, and plants, unable to photosynthesize, would have shrivelled too. Perhaps, if the dust cleared through rainfall during those first grim months, they did not all succumb, and as light and warmth returned, they became restored. Insects could rest dormant and perhaps fishes underwater escaped some of the worst effects.

A study in 2018 of the Cretaceous–Palaeogene rock section at El Kef in Tunisia has shown that temperatures rose by 5 degrees for about 100,000 years, immediately after the impact. This was determined by measurements of oxygen isotopes from fish bones buried in the rocks, and the high temperatures persisted for about 3 metres (10 feet) of the section. The increase in temperature was probably driven by additional carbon dioxide in the atmosphere produced by melting of limestones by the impact and by the forest fires that followed soon after.

Such a modest rise in temperature would have driven some species from tropical areas during the initial stages of recovery from the crisis, but the change was not large enough in itself to contribute to a great deal of further extinction. All the evidence suggests, then, that the crisis was short and brutal, and that there were persistent environmental changes induced by the impact. But the Earth and life recovered amazingly fast, in geological terms, say within 100,000 years of those elevated temperatures, and it is worth looking in more detail at the fates of two of the winning groups: the birds and mammals.

How did birds survive the mass extinction?

Birds survived the cataclysm 66 million years ago, and a 2018 study has shown what happened. Like mammals, birds had already evolved for some time in the Mesozoic, but identifiable birds are known only from the Late Jurassic onwards, say from 155 million years ago. These first birds include the famous 'ancient bird' *Archaeopteryx* from the Solnhofen lagoons of southern Germany (see p. 111), as well as other examples from rocks of similar, or even slightly older, provenance from north China. In fact, the crossover from small feathered theropod dinosaurs to birds was subtle, so deciding which was the last dinosaur and which was the first bird on that evolutionary branch is a merely semantic exercise.

It used to be thought that birds had survived the mass extinction with minimal damage – a few ancient lines had died out, but birds survived and then diversified rapidly to reach eventually the 11,000 living species. This was supposed to confirm their great adaptability, hence their modern success, while emphasizing the contrast between them and their relatives the dinosaurs, which all died out. However, the story seems different now.

Close study of the bird fossil record, coupled with exploration of modern bird evolution based on genomic analysis, shows that only five species of birds crossed the Cretaceous–Palaeogene boundary. In fact, it seems that the whole fluttering and twittering multitude of birds came close to annihilation. So, what happened?

As we saw earlier, the Jehol Beds of north China have produced thousands of astonishing specimens of Cretaceous birds, showing us a far higher diversity than we had ever imagined. Four main groups of birds reached the end of the Cretaceous and were killed off by the asteroid strike. These are the enantiornithines, a diverse group of eighty species of flying birds, whose diets included hard-shelled prey caught in

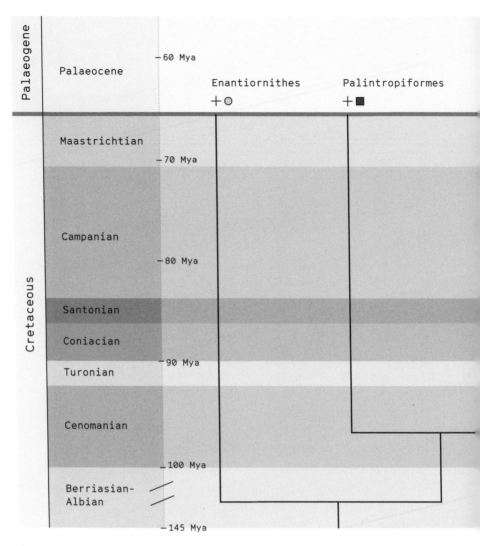

Evolution of birds through the Cretaceous–
Palaeogene interval, showing that mainly
ground-dwellers survived.

the water or by probing in mud, or vertebrates and arthropods, possibly snatched from the trees. The second group, the palintropiforms, comprise only two or three species from Mongolia. The ichthyornithines were fish-eating, flying birds with powerful toothy jaws, known from the shores of the Western Interior Seaway of North America. Finally, the hesperornithiforms, also known mainly from the same localities in North America, were generally large, flightless, diving, fish-hunting birds.

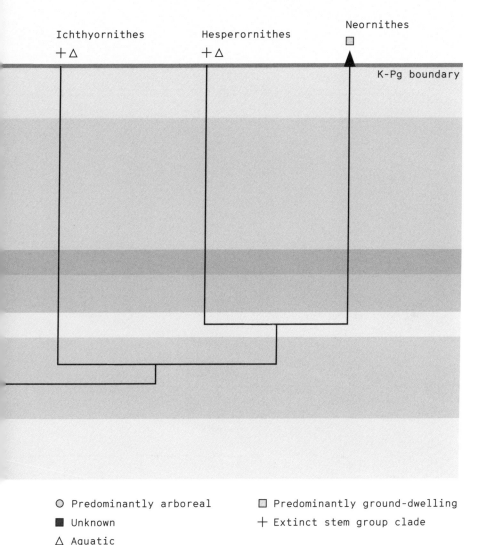

Ichthyornithes
+△

Hesperornithes
+△

Neornithes
□

K-Pg boundary

○ Predominantly arboreal   □ Predominantly ground-dwelling
■ Unknown   + Extinct stem group clade
△ Aquatic

The only survivors were a few species of ground-dwelling ancestors of ducks and chickens. This could be explained simply by saying they were lucky – but a 2018 study suggests something else. Dan Field, an enthusiastic young palaeo-ornithologist, worked with colleagues on a large-scale ecological study of birds. Using a phylogenetic tree of modern species, they annotated which forms are tree-dwellers, such as robins, owls, swifts, and parrots, and which live mainly on the ground, such as

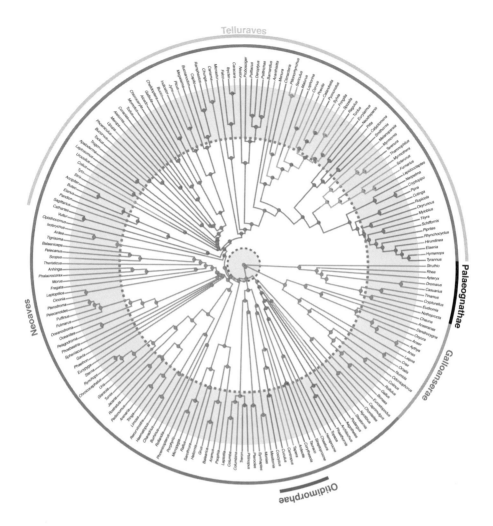

Phylogeny of modern birds, showing that the
ancestors of most were ground-dwellers.

ostriches, ducks, chickens, rails, shorebirds, and gulls. Tracking back to the
latest Cretaceous, they found that the ancestors of all modern birds had
been ground-dwellers at the time of the asteroid hit.

All the tree-dwellers, together with the unfortunate ichthyornithines
and hesperornithiforms, died out. Field and colleagues coupled this
finding with evidence from the fossil records of plants and pollen, which
showed that forests had been devastated for about a thousand years by
the consequences of the asteroid strike. The blotting out of the sun would
have halted photosynthesis and generally killed plants, while acid rain
and other dire consequences of the impact probably flattened forests.

Birds and other animals that lived in the complex ecosystems in and around trees would have been left homeless.

Then, in the Palaeogene, Field and colleagues noted that the first birds of modern aspect to emerge were largely ground-dwellers, scampering around after insects or feeding on marine life on the shore. They could tell the broad mode of life of the fossils by looking at the relative lengths of portions of the leg – tree-dwellers tend to have quite long femurs, whereas ground runners have shorter femurs, and the parts of the leg below the knee are longer. True tree-dwelling took some time to re-emerge, and many different lines of birds independently adapted back to tree life when the opportunity arose.

The survival of birds through the mass extinction was, one might say, a close-run thing then. What about the mammals?

How did mammals take over from dinosaurs?

The classic story is that after the mass extinction 66 million years ago, the mammals took over – which indeed they did. The asteroid impact was tough for the dinosaurs, as well as for the pterosaurs, marine reptiles, ammonites, belemnites, and many other groups, but it was good for our ancestors, and therefore ultimately for us. Mammals had, in fact, originated at the same time as the dinosaurs, in the Triassic, and yet most of them remained small and nocturnal throughout the Mesozoic. How can we be sure that it was the removal of dinosaurs from ecosystems that enabled mammals to diversify in the Palaeogene?

The answer comes from numerical modelling. Graham Slater of the University of Chicago ran tests in a 2013 paper. First, he constructed a dated phylogenetic tree of the evolution of early mammals through the Mesozoic and Palaeogene. He then tried fitting all kinds of evolutionary models to the data, and found that the best one in terms of its ability to explain the data was what he called the 'release and radiate' model – this model described the evolution of mammals as having been held back by the dinosaurs, and then, when the dinosaurs disappeared, they were released from the negative pressure and radiated explosively. The other models, including various random patterns, a driven trend, or simply extinction of the dinosaurs with no ecological reaction from mammals.

So, the old assumption that mammals were held back by the dinosaurs has been confirmed by a smart computational model – when competition from the large, daytime dinosaurs had disappeared,

crown Eutheria

The explosive evolution of mammals after the Cretaceous–
Palaeogene event, 66 million years ago.

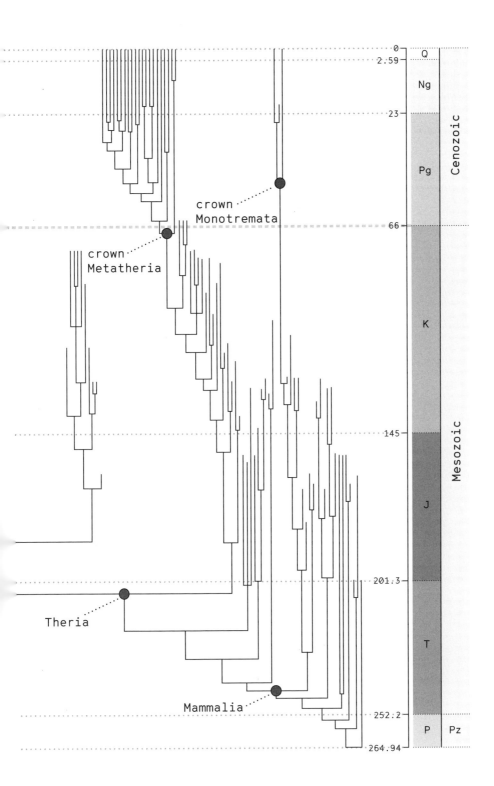

mammals could evolve diurnal forms and larger forms, covering the whole range of habitats. Indeed, their evolution was particularly rapid in the early Palaeogene, which is striking because they had been around for 170 million years of the Mesozoic. Ten million years after the mass extinction, mammals had evolved all modern groups, ranging from small shrew-like creatures to flying bats and monster whales, from large, armoured herbivores to massive-skulled predators, and even monkey-like tree-dwellers, including our direct ancestors.

The evolution of mammals in the past 66 million years has seen continuing expansion worldwide. As temperatures zig-zagged up and down, but mainly down, habitats changed profoundly. Most notable was the spread of grasslands about 30 million years ago. As climates cooled, rainfall patterns became less tropical and more temperate, and the centres of continents became dry. The forests drew back, and great prairies spread over much of North America and South America, as well as central Africa and Asia. Whereas mammals until then had been adapted to living somewhat secretive lives in the forests, the new grasslands provided opportunities. The ancestors of horses and rhinos on the one hand and cattle and deer on the other crept out of the edges of the forests, and some began eating the burgeoning new grasses.

This was tough food, because most grasses contain silica to protect themselves from grazing, so the herbivores had to evolve deeper and tougher teeth that could grind the grit. Further, the early grassland herbivores were vulnerable out in the open. Their small size, which had worked to protect them in the forests, was no longer an advantage. Successful species evolved to be taller, with long thin legs so they could peer around for predators, but also run off at speed when threatened. Famously, horses evolved from animals the size of a Labrador dog to the modern size at this time, and the same goes for most other successful herbivores. The predators, too, adapted to running rather than climbing.

One view of early human evolution posits a similar shift from forest to plain. Our nearest relatives, chimpanzees and gorillas, stayed in the African forests where they still reside, but our ancestors apparently stepped out onto the hot grasslands of central and eastern Africa 5 million years ago. We owe our upright posture to the need of these early ape-persons to peep about over the tall grass in search of lions and cheetahs, and then to carry food and tools in their free hands.

Study of the end-Cretaceous mass extinction has been a helter-skelter of new ideas. It's also a great example of how speculation has turned into science, and then told us things we never expected. The brilliance of the insights by Luis Alvarez and his team, when they proposed their model for asteroid impact and global devastation back in 1980, has been borne out by thousands of investigations since. What they predicted has been found, and few now doubt the strength of the evidence in favour of impact – the Chicxulub crater in Mexico, the tsunami beds and evidence of melt beads all round the proto-Caribbean, and the worldwide spread of fine dust with iridium enrichment.

Palaeobiological work has also massively refined our picture of the sequence of events of the mass extinction. It seems that dinosaurs were already in decline, slowly, over their last 40 million years on Earth, but in the end the impact and subsequent cold and darkness finished them off at a stroke. Birds and mammals also suffered, and only certain species struggled through the grim environmental conditions. These surviving birds and mammals then evolved explosively when the world settled down after the crisis, and they form the basis of modern terrestrial ecosystems.

There is much more to be done in focusing ever closer on the event, to work out the step-by-step deterioration of environments. Also, we cannot yet fully integrate the devastation caused by the asteroid strike with that caused by the longer-running Deccan Traps eruptions in India. Then, the nature of evolution before, during, and after the crisis is yet to be explored fully. Recent studies of dinosaur decline and selective extinction and then diversification of birds and mammals are resolving long-running questions. The fact that these events 66 million years ago still resonate in the structure of modern ecosystems adds urgency to these further studies.

Afterword

During the past forty years, we have seen the transformation of palaeontology into science. This has involved expansion of the area of testable science, and reduction of the area of speculation. Arguably, the area of speculation is infinite, and probably grows as people ask more questions. Nonetheless, I have documented how major questions about function (What did that dinosaur eat? How fast could it run? What colour was it? How fast did it grow?) can now all be answered by testable means.

What is science?

Philosophers have long debated the definition of science. One thing for sure is that there is more to science than simply mathematics, where proof is possible. In all other science, proof is never possible, merely disproof. This leads to massive misunderstanding, especially by flat-earthers, climate-change deniers, creationists and others – they like to argue that science is about facts, and think that if they can dislodge a fact, the whole science fails.

In reality, natural science is about hypotheses and theories. I can have many hypotheses about why the dinosaurs died out (see Appendix) or why sauropods were so large, and some of these modify into theories when there is a sufficient body of coherent evidence. For example, the Alvarez model for catastrophic impact at the end of the Cretaceous and its role in killing the dinosaurs is a theory with very good evidence. It cannot be proved, but it could easily be disproved. Especially when Luis Alvarez and his team established the theory in 1980, the actual evidence was modest – essentially the iridium spike from two localities, Gubbio and Stevns Klint. Year by year, evidence accumulated that confirmed their suggestion, and nothing was found to refute it. If it had been shown that the iridium spikes were all of different geological ages, or the iridium spike only occurred in those two localities, the theory would have been refuted. Then, in 1991, scientists found the impact crater itself, which gave good, clinching evidence for the Alvarez theory. The exact role of the asteroid impact in killing the dinosaurs, and whether it did so alone or with additional pressure from changing climates and the eruption of the

Deccan Traps, are still matters of some debate. But the theory is the most convincing one yet.

This is the principle of the hypothetico-deductive method in science, as enunciated by Karl Popper in 1934. He could see the method extended to the historical sciences, but was a little unsure how. At first, it might seem to be forever impossible to discuss historical events, whether in archaeology, palaeontology, or geology, in a scientific manner. However, in this book, we have explored some examples where it works.

Theory and criticism

Scientists and non-scientists can often misunderstand the role of criticism. Of course, it is the duty of all observers to correct mistaken facts – so palaeontologists, and an army of online commentators, are quick to pick up mistakes in dating fossils, identifying specimens, measuring dimensions, reporting specific anatomical features and the like. Charles Darwin, who was right about so much, said this: 'False facts are highly injurious to the progress of science, for they often endure long; but false views, if supported by some evidence, do little harm, for every one takes a salutary pleasure in proving their falseness.'

Criticism of hypotheses and theories is another matter. Indeed, as Darwin said, scientists do not hold back from criticizing the theories of others. But the critic cannot simply ridicule a particular theory and step back. It's not like politics. The critic must present a more convincing theory that explains the data *better* than the theory he or she is criticizing. The business of being a scientist is rigorous, and the argument has to consider all evidence, and weigh up alternative hypotheses in an even-handed way.

This brings us to the idea of a 'theory'. It's true in English, and in many other languages, that the term 'theory' has two meanings. In common speech, a 'theory' can be a notion, such as 'my theory is that we'll have sausages for supper tonight...and it's my neighbour's dog that is messing the garden'. These are small-scale deductions, theories of a sort. In science, a theory is a model of how the world works, such as gravity or evolution (big theories) or the end-Cretaceous impact or the replacement of dinosaurs by mammals through competitive release (smaller theories). The evidence lines up, they have been repeatedly tested, they are robust, and there is no better theory around. As for the sausages or the dog, who knows? There could be a million alternative outcomes or explanations.

So, climate-change deniers and creationists and, in their time, the smoking-as-killer deniers, play with the word 'theory'. 'Oh, it's just a guess', they say. Their alternative perspectives, however, do not stand up to the evidence. That dinosaurs existed is a theory. So is gravity, and so is the germ theory of disease, and we're prepared to risk our lives by flying in an aeroplane or going under the surgeon's knife based on those theories – because they are real theories that have been stress-tested.

The transformation of dinosaur palaeobiology from speculation to science

In this book, we have been on a journey from 1980 to the present day, and we have stopped off at some of the debates and controversies. In particular, we have looked at those cases where the experts in the past could only really speculate and give an opinion, and how those fields have been transformed into science.

We saw how the evolutionary tree and classification of dinosaurs were transformed from 1984 onwards by the application of cladistic methods – and how the debates were reignited in 2017 with a radical new tree of dinosaurs. We have seen how those trees are the basis for the application of new methods in the study of evolution to look for fast and slow rates of change, and even the kinds of models of evolution, including the tension, triggered by one of our papers in 2016, about whether or not dinosaurs were somehow on the way out long before the asteroid struck.

Indeed, the idea that dinosaurs were exterminated by the consequences of the impact of a huge asteroid, proposed in 1980, and tested hard ever since, has led to the most extraordinary revitalization of many fields of science.

And, as we have seen, it's not just evolutionary trees and events that have been rethought, but also palaeobiology. New engineering approaches have revolutionized the ways in which dinosaurologists investigate feeding and locomotion, and close observations of bone histology allow us to make some conclusions about growth and physiology. Most remarkable of all, arguably, was the discovery of the colours and patterns of dinosaur feathers in 2010, and the implications of that discovery for our understanding of sexual behaviour and perhaps the role of sexual selection in dinosaur evolution.

What next?

What next? Ten years ago, I would have said that for sure we would never know the colour of dinosaurs. Now we do – well, some at least, and based on a solid line of reasoning. So, we now have the melanosome theory for ancient colour. Perhaps we don't know the noises dinosaurs made. Also, it seems unlikely we can ever find dinosaur DNA or clone a dinosaur... Still, should we say 'never'?

Advances in methods of chemical analysis, and experimental approaches, are improving our knowledge of which organic molecules are capable of surviving the rigours of fossilization. Some, such as melanin, can survive, and other dinosaur molecules such as proteins might provide sequence information that could be used to test relationships in evolutionary trees. We have only just begun CT scanning and engineering analysis of fossils, and much new information about how dinosaurs moved and fed will emerge. Another growth field is number-crunching of evolutionary patterns and processes on supertrees, and new methods will lead to new discoveries. Microscopic studies of bone will tell us more about dinosaur growth and sex, and tooth-wear patterns may confirm ideas about diet.

Scientists and the public are understandably fascinated by the extinction of the dinosaurs and how and why it happened. Surely, of equal fascination is the rather obscure origin of the dinosaurs. We have seen how new fossils have pushed the date of origin back from 230 million years ago to 245 million years ago, and – who knows? – future researchers may push it back even further. We have also seen that in their origin dinosaurs were playing out a complex ecological story, and we are only now getting to grips with it. As we saw, there's a fundamental tension between competitive and opportunistic models – were dinosaurs driving their competitors to the wall and showing their progressive qualities, or were they the lucky beneficiaries of a series of grim environmental catastrophes? These are core questions about life on Earth, rates of change, and the role of climate change in driving evolution – all very current issues key to understanding the future fate of biodiversity.

In his 1998 book *What Remains to be Discovered*, John Maddox, veteran science commentator and editor, was quite clear that much remained to be discovered in science, and each new discovery leads to fresh questions. The same is undoubtedly true of palaeobiology and the science of dinosaurs. We eagerly await the wonders that will be brought to the table by new generations of researchers.

Extinction Hypotheses

List of published hypotheses for the extinction of the dinosaurs. The date of first publication of each idea is given, where possible. Most of these ideas would be rated as 'nonsense', because there is no evidence. I highlight those ideas for which there is some evidence, and mark two that are plausible contributors to the mass-extinction theory in **bold**. These ideas are fully described and referenced in a paper I wrote years ago (Benton, M.J. 1990. Scientific methodologies in collision: the history of the study of the extinction of the dinosaurs. *Evolutionary Biology*, 24, 371–424).

1. Biological causes

A.   'Medical problems'

A1. Metabolic disorders
  01. Slipped vertebral discs
  02. Malfunction or imbalance of hormone systems
      01. Overactivity of pituitary gland and excessive growth of bones and cartilage [1917]
      02. Malfunction of pituitary gland leading to excess growth of debilitating horns, spines, and frills [1910]
      03. Imbalances of vasotocin and oestrogen levels leading to pathological thinning of egg shells [1979]
  03. Diminution of sexual activity [1917]
  04. Cataract blindness [1982]
  05. Disease: caries, arthritis, fractures, and infections reached a maximum in Late Cretaceous reptiles [1923]
  06. Epidemics
  07. Parasites
  08. AIDS caused by increasing promiscuity [1986]
  09. Change in ratio of DNA to cell nucleus

A2. Mental disorders
      01. Dwindling brain and consequent stupidity [1939]
      02. Absence of consciousness and ability to modify behaviour [1979]
      03. Development of psychotic suicidal factors
      04. *Paleoweltschmerz*: boredom with the ancient world

A3. Genetic disorders: excessive mutation rate induced by high levels of cosmic rays and/or ultraviolet light, leading to small population size burdened by a high genetic load, and consequent vulnerability to environmental shock [1987]

B.   Racial senility
   01. Evolutionary drift into senescent overspecialization, as evinced in gigantism and spinescence (e.g., loss of teeth, and 'degenerate form') [1910]
   02. Racial old age [1964, Will Cuppy: 'the Age of Reptiles ended because it had gone on long enough and it was all a mistake in the first place']

C.   Biotic interactions

C1. Competition with other animals
   01. Competition with the mammals – invasion of North America by Asian mammals [1922]
   02. Competition with caterpillars, which ate all the plants [1962]

C2. Predation
   01. Overkill capacity by predators (the theropods ate themselves out of existence)
   02. Egg-eating by mammals, which reduced hatching success of the young [1925]

C3. Floral changes
   01. Spread of flowering plants and reduction in availability of conifers, ferns, etc. This led to a reduction of fern oils, and to lingering death by terminal constipation [1964]
   02. Floral change and loss of marsh vegetation [1922]
   03. Floral change and increase in forests [1981]
   04. Reduction in availability of plant food as a whole
   05. Presence of poisonous tannins and alkaloids in the flowering plants [1976]
   06. Presence of other poisons in plants
   07. Lack of calcium and other necessary minerals in plants
   08. Rise of flowering plants, and of their pollen, leading to extinction of dinosaurs by hay fever [1983]

2. Physical environmental causes

D.   Terrestrial explanations

D1. Climatic change
   01. Climate became too hot as a result of high levels of carbon dioxide
       in the atmosphere, and the 'greenhouse effect'; extinction was
       caused by the high temperature and increased aridity [1946],
       which either inhibited spermatogenesis [1945], unbalanced the
       male:female ratio of hatchlings [1982], killed off juveniles [1949],
       or led to overheating in summer, especially if the dinosaurs were
       warm-blooded [1978]
   02. Climate became too cold, and this led to extinction because it
       was too cold for embryonic development [1929], because the
       endothermic dinosaurs lacked insulation and could not maintain
       a constant body temperature [1965], and they were also too large
       to hibernate [1967], or, even if they were warm-blooded, the cold
       winter temperatures finished them off [1973]
   03. Climate became too dry [1946]
   04. Climate became too wet
   05. Reduction in climatic equability and increase in seasonality [1968]

D2. Atmospheric change
   01. Changes in the pressure or composition of the atmosphere
       (e.g. excessive amounts of oxygen from photosynthesis) [1957]
   02. High levels of atmospheric oxygen, leading to fires following
       an extraterrestrial impact [1987]
   03. Low levels of carbon dioxide, removing the 'breathing stimulus'
       of dinosaurs [1942]
   04. Excessively high levels of carbon dioxide in the atmosphere and
       asphyxiation of dinosaur embryos in the eggs [1978]
   05. Extensive volcanism and the production of volcanic dust
   06. Poisoning by selenium from volcanic lava and dust [1967]
   07. Toxic substances in the air, possibly produced from volcanoes,
       which caused thinning of dinosaur egg shells [1972]

D3. Oceanic and topographic change
    01. Sea level fall and drying of continental interiors [1964]
    02. Lowering of global sea level leading to dinosaur extinction, on the assumption that they were underwater organisms [1949]
    03. Floods
    04. Mountain building, for example, the Laramides in North America [1921]
    05. Drainage of swamp and lake habitats [1939]
    06. Stagnant oceans caused by high levels of carbon dioxide [1983]
    07. Loss of oxygen on the sea bed at start of sea level advance [1984]
    08. Spillover of Arctic water (fresh) from its formerly enclosed condition into the oceans, which led to reduced temperatures worldwide [1978]
    09. Reduced topographic relief, and reduction in terrestrial habitats [1968]

D4. Other terrestrial catastrophes
    01. Eruption of the Deccan Traps [1982]
    02. Fluctuation of gravitational constants
    03. Shift of the Earth's rotational poles
    04. Extraction of the moon from the Pacific Basin and consequent world perturbation
    05. Poisoning by uranium sucked up from the soil [1984]

E. Extraterrestrial explanations
    01. Entropy; increasing chaos in the universe and hence loss of large organized life forms
    02. Sunspots
    03. Cosmic radiation and high levels of ultraviolet radiation [1928]
    04. Destruction of the ozone layer by solar flares, letting in ultraviolet radiation [1954]
    05. Ionizing radiation [1968]
    06. Electromagnetic radiation and cosmic rays from the explosion of a nearby supernova [1971]
    07. Interstellar dust cloud [1984]
    08. Flash heating of atmosphere by entry of meteorite [1956]
    09. Oscillations about the galactic plane [1970]
    10. Impact of an asteroid, a comet, or comet showers (Luis Alvarez and colleagues, 1980)

Further Reading

Author's note

Publications marked with an asterisk (*) provide easy introductions to the topic, and should be consulted first.

Some key publications described in the book are listed below for the dedicated scholar, or for anyone who wants to check up on what I have said. Brief comments after particular papers clarify any further details.

All quotations from individuals were provided directly to me by email, unless otherwise indicated in the text.

I also add a list of current dinosaur books that are all original and worth reading.

Introduction
How Scientific Discoveries Are Made   (pp. 8–20)

Benton, M. J. 2015. *When Life Nearly Died*. 2nd edition. Thames & Hudson, London and New York

*Magee, B. 1974. *Popper*. Routledge, London
An excellent, short introduction to Karl Popper's writings on philosophy, including the hypothetico-deductive method in science and his thoughts on the historical sciences.

Popper, K. R. 1934. *Logik der Forschung*. Mohr Siebeck, Tübingen [First English-language edition, *The logic of scientific discovery*, published by Routledge, London.]

Rhodes, F. H. T., Zim, H. S., and Shaffer, P. R. 1962. *Fossils, a guide to prehistoric life*. Golden Nature Guides, Golden Press, New York; Hamlyn, London

Witmer, L. M. 1995. The extant phylogenetic bracket and the importance of reconstructing soft tissues in fossils. In J. J. Thomason (ed.), *Functional morphology in vertebrate paleontology*, Cambridge University Press, Cambridge and New York, pp. 19–33

Benton, M. J. 1983. Dinosaur success in the Triassic: A noncompetitive ecological model. *Quarterly Review of Biology* 58, 29–55
My original paper that challenged the Romer-Colbert-Charig model for dinosaur origins by competitive relay.

Benton, M. J., Bernardi, M., and Kinsella, C. 2018. The Carnian Pluvial Episode and the origin of dinosaurs. *Journal of the Geological Society* 175 (6), 1019

*Benton, M. J., Forth, J., and Langer, M. C. 2014. Models for the rise of the dinosaurs. *Current Biology* 24, R87–R95
A brief, but slightly technical, introduction to how we use numerical methods to explore the origin of dinosaurs.

Bernardi, M., Gianolla, P., Petti, F. M., Mietto, P., and Benton, M. J. 2018. Dinosaur diversification linked with the Carnian Pluvial Episode. *Nature Communications* 9, 1499: https://www.nature.com/articles/s41467-018-03996-1

Brusatte, S. L., Benton, M. J., Ruta, M., and Lloyd, G. T. 2008. Superiority, competition, and opportunism in the evolutionary radiation of dinosaurs. *Science* 321, 1485–88

*Brusatte, S. L., Nesbitt, S. J., Irmis, R. B., Butler, R. J., Benton, M. J., and Norell, M. A. 2010. The origin and early radiation of dinosaurs. *Earth-Science Reviews* 101, 68–100
A broad overview of all aspects of the Triassic and the origin of the dinosaurs.

Brusatte, S. L., Niedźwiedzki, G., and Butler, R. J. 2011. Footprints pull origin and diversification of dinosaur stem lineage deep into Early Triassic. *Proceedings of the Royal Society B* 278, 1107–13

Dal Corso, J. et al. 2012. Discovery of a major negative $\delta^{13}C$ spike in the Carnian (Late Triassic) linked to the eruption of Wrangellia flood basalts. *Geology* 40, 79–82

Dzik, J. 2003. A beaked herbivorous archosaur with dinosaur affinities from the early Late Triassic of Poland. *Journal of Vertebrate Paleontology* 23, 556–74

Nesbitt, S. J., Sidor, C. A., Irmis, R. B., Angielczyk, K. D., Smith, R. M. H., and Tsuji, L. A. 2010. Ecologically distinct dinosaurian sister group shows early diversification of Ornithodira. *Nature* 464, 95–98
Announcement of the Middle Triassic silesaurid, which confirms an early date for dinosaur origins.

Simms, M. J., and Ruffell, A. H. 1989. Synchroneity of climatic change and extinctions in the late Triassic. *Geology* 17, 265–68

Sookias, R. B., Butler, R. J., and Benson, R. B. J. 2012. Rise of dinosaurs reveals major body-size transitions are driven by passive processes of trait evolution. *Proceedings of the Royal Society B* 279, 2180–87

Chapter 2
Making the Tree

(pp. 52–84)

Bakker, R. T., and Galton, P. M. 1974. Dinosaur monophyly and a new class of vertebrates. *Nature* 248, 168–72

Baron, M. G., Norman, D. B., and Barrett, P. M. 2017. A new hypothesis of dinosaur relationships and early dinosaur evolution. *Nature* 543, 501–6

Benton, M. J. 1984. The relationships and early evolution of the Diapsida. *Symposium of the Zoological Society of London* 52, 575–96

*Brusatte, S. L. 2012. *Dinosaur paleobiology*. Wiley, New York and Oxford
The best student textbook about dinosaurs.

Gauthier, J. 1986. Saurischian monophyly and the origin of birds. *Memoirs of the California Academy of Science* 8, 1–55

*Gee, H. 2008. *Deep time: Cladistics, the revolution in evolution*. Fourth Estate, London
A great introduction to cladistics and all the squabbles.

Hennig, W. 1950. *Grundzüge einer Theorie der phylogenetischen Systematik*. Deutscher Zentralverlag, Berlin

Hennig, W. 1966. *Phylogenetic systematics*, translated by D. Davis and R. Zangerl. University of Illinois Press, Urbana

Lloyd, G. T., Davis, K. E., Pisani, D., Tarver, J. E., Ruta, M., Sakamoto, M., Hone, D. W. E., Jennings, R., and Benton, M. J. 2008. Dinosaurs and the Cretaceous Terrestrial Revolution. *Proceedings of the Royal Society, Series B* 275, 2483–90
Our second dinosaur supertree.

*Naish, D., and Barrett, P. 2016. *Dinosaurs: How they lived and evolved*. Natural History Museum, London; Smithsonian Books, Washington DC
An excellent and colourful introduction to the latest dinosaur finds.

Norman, D. B. 1984. A systematic reappraisal of the reptile order Ornithischia. In W.-E. Reif and F. Westphal (eds), *Third Symposium on Mesozoic terrestrial ecosystems, short papers*, Attempto Verlag, Tübingen, pp. 157–62

Owen, R. 1842. Report on British fossil reptiles. Part II. *Report of the Eleventh Meeting of the British Association for the Advancement of Science; held at Plymouth in July 1841* 60–204
The classic paper in which Richard Owen named the Dinosauria (see p. 103).

Pisani, D., Yates, A. M., Langer, M. C., and Benton, M. J. 2002. A genus-level supertree of the Dinosauria. *Proceedings of the Royal Society B* 269, 915–21
Our first dinosaur supertree.

Seeley, H. G. 1887. On the classification of the fossil animals commonly named Dinosauria. *Proceedings of the Royal Society, London* 43, 165–71

Sereno, P. C. 1986. Phylogeny of the bird-hipped dinosaurs (order Ornithischia). *National Geographic Research* 2, 234–56

Sweetman, S. C. 2016. A comparison of Barremian–early Aptian vertebrate assemblages from the Jehol Group, north-east China and the Wealden Group, southern Britain: The value of microvertebrate studies in adverse preservational settings. *Palaeobiodiversity and Palaeoenvironments* 96, 149–68

Chapter 3
Digging Up Dinosaurs

*Benton, M. J., Schouten, R., Drewitt, E. J. A., and Viegas, P. 2012. The Bristol Dinosaur Project. *Proceedings of the Geologists' Association* 123, 210–25
The full story of *Thecodontosaurus* and how we have built an educational and engagement programme around this dinosaur.

*Currie, P. J., and Koppelhus, E. B. (eds). 2005. *Dinosaur Provincial Park: A spectacular ancient ecosystem revealed.* Indiana University Press, Bloomington
The whole story – everything about the Dinosaur Park Formation: its geology, plants, animals, and dinosaurs.

Bakker, R. T. 1972. Anatomical and ecological evidence of endothermy in dinosaurs. *Nature* 238, 81–85
The paper that kicked off the 'warm-blooded dinosaurs' debate.

*Bakker, R. T. 1986. *The dinosaur heresies: New theories unlocking the mystery of the dinosaurs and their extinction.* W. Morrow, New York; Longman, Harlow
The title says it all.

Bakker, R. T., and Galton, P. M. 1974. Dinosaur monophyly and a new class of vertebrates. *Nature* 248, 168–72

Benton, M. J. 1979. Ectothermy and the success of the dinosaurs. *Evolution* 33, 983–97
My paper about the 'warm-blooded dinosaurs' controversy.

*Benton, M. J., Zhou, Z., Orr, P. J., Zhang F., and Kearns, S. L. 2008. The remarkable fossils from the Early Cretaceous Jehol Biota of China and how they have changed our knowledge of Mesozoic life. *Proceedings of the Geologists' Association* 119, 209–28
An overview of the Chinese feathered dinosaur fossils and their occurrence.

Chen, P., Dong, Z., and Zhen, S. 1998. An exceptionally well-preserved theropod dinosaur from the Yixian Formation of China. *Nature* 391, 147–52
The first description of a feathered dinosaur, in English, presenting *Sinosauropteryx* to the world.

*Chiappe, L. M., and Meng, Q. J. 2016. *Birds of stone: Chinese avian fossils from the age of dinosaurs.* Johns Hopkins University Press, Pittsburgh
All the amazing fossil birds.

Colbert, E. H., Cowles, R. B., and Bogert, C. M. 1946. Temperature tolerances in the American alligator and their bearing on the habits, evolution, and extinction of the dinosaurs. *Bulletin of the American Museum of Natural History* 86, 327–74
Those great experiments where Colbert and colleagues found that being large can help you keep a more or less constant body temperature.

Huxley, T. H. 1870. Further evidence of the affinity between the dinosaurian reptiles and birds. *Quarterly Journal of the Geological Society of London* 26, 12–31
Huxley shows that dinosaurs and birds share much of their anatomy.

Jerison, H. J. 1969. Brain evolution and dinosaur brains. *American Naturalist* 103, 575–88

Knell, R. J., and Sampson, S. 2011. Bizarre structures in dinosaurs: Species recognition or sexual selection? A response to Padian and Horner. *Journal of Zoology* 283, 18–22
Argues that horns and crests are for sexual display, not species recognition.

Li, Q., Gao, K.-Q., Vinther, J., Shawkey, M. D., Clarke, J. A., D'Alba, L., Meng, Q., Briggs, D. E. G., Miao, L., and Prum, R. O. 2010. Plumage color patterns of an extinct dinosaur. *Science* 327, 1369–72
The Yale group show the colours and patterns of feathers in the Jurassic dinosaur *Anchiornis*.

*Long, J., and Schouten, P. 2009. *Feathered dinosaurs: The origin of birds*. Oxford University Press, Oxford and New York
Spectacular illustrations and the importance of the new fossils from China.

Ostrom, J. H. 1969. Osteology of *Deinonychus antirrhopus*, an unusual theropod from the Lower Cretaceous of Montana. *Bulletin, Peabody Museum of Natural History* 30, 1–165
The classic description of *Deinonychus*, and the paper that showed birds evolved from dinosaurs.

Padian, K., and Horner, J. 2011. The evolution of 'bizarre structures' in dinosaurs: Biomechanics, sexual selection, social selection, or species recognition? *Journal of Zoology* 283, 3–17
Argues that horns and crests are for species recognition, not sexual display.

Vinther, J., Briggs, D. E. G., Prum, R. O., and Saranathan, V. 2008. The colour of fossil feathers. *Biology Letters* 4, 522–25
The case for melanosomes in fossil feathers.

Xing, L., McKellar, R. C., Xu, X., Li, G., Bai, M., Persons, W. S. IV, Miyashita, T., Benton, M. J., Zhang, J. P., Wolfe, A. P., Yi, Q. R., Tseng, K. W., Ran, H., and Currie, P. J. 2016. A feathered dinosaur tail with primitive plumage trapped in mid-Cretaceous amber. *Current Biology* 26, 3352–60

Zhang, F., Kearns, S. L, Orr, P. J., Benton, M. J., Zhou, Z., Johnson, D., Xu, X., and Wang, X. 2010. Fossilized melanosomes and the colour of Cretaceous dinosaurs and birds. *Nature* 463, 1075–78
Our paper in which we show *Sinosauropteryx* had a ginger and white stripy tail.

Quotations from Mary Schweitzer come from M. Schweitzer and T. Staedter, The real Jurassic Park. *Earth* June 1997, 55–57

*Briggs, D. E. G., and Summons, R. E. 2014. Ancient biomolecules: Their origins, fossilization, and role in revealing the history of life. *BioEssays* 36, 482–90
A clear account of which biological molecules are likely to survive for millions of years, and which are not.

Buckley, M., Warwood, S., van Dongen, B., Kitchener, A. C., and Manning, P. L. 2017. A fossil protein chimera: Difficulties in discriminating dinosaur peptide sequences from modern cross-contamination. *Proceedings of the Royal Society B* 284, 20170544
Rejection of reported blood vessels in dinosaur bone as bacterial biofilms.

Burroughs, E. R. 1918. *The Land that Time Forgot*. A. C. McClurg, Chicago

Cano, R. J., Poinar, H. N., Pieniazek, N. J., Acra, A., and Poinar, G. O., Jr. 1993. Amplification and sequencing of DNA from a 120–135-million-year-old weevil. *Nature* 363, 536–38

Cano, R. J., Poinar, H. N., Roubik, D. W., and Poinar, G. O., Jr. 1992. Enzymatic amplification and nucleotide sequencing of portions of the 18s rRNA gene of the bee *Proplebeia dominicana* (Apidae: Hymenoptera) isolated from 25–40-million-year-old Dominican amber. *Medical Science Research* 20, 619–22

Chinsamy, A., Chiappe, L. M., Marugan-Lobon, J., Gao, C. L., and Zhang, F. J. 2013. Gender identification of the Mesozoic bird *Confuciusornis sanctus*. *Nature Communications* 4, 1381
Medullary bone identifies a female fossil bird.

*Crichton, M. 1990. *Jurassic Park*. Alfred A. Knopf, New York
The book that started it all.

Doyle, A. C. 1912. *The Lost World*. Hodder & Stoughton, London

Kaye, T. G., Gaugler, G., and Sawlowicz, Z. 2008. Dinosaurian soft tissues interpreted as bacterial biofilms. *PLoS ONE* 3, e2808
Rejection of reported blood vessels in dinosaur bone as bacterial biofilms.

Kupferschmidt, K. 2014. Can cloning revive Spain's extinct mountain goat? *Science* 344, 137–38

Lindahl, T. 1993. Instability and decay of the primary structure of DNA. *Nature* 362, 709–15
Evidence, from the start, that ancient DNA was unlikely to survive for millions of years.

Muyzer, G., Sandberg, P., Knapen, M. H. J., Vermeer, C., Collins, M., and Westbroek, P. 1992. Preservation of the bone protein osteocalcin in dinosaurs. *Geology* 20, 871–74

O'Connor, R. E., Romanov, M. N., Kiazim, L. G., Barrett, P. M., Farré, M., Damas, J., Ferguson-Smith, M., Valenzuela, N., Larkin, D. M., and Griffin, D. K. 2018. Reconstruction of the diapsid ancestral genome permits chromosome evolution tracing in avian and non-avian dinosaurs. *Nature Communications* 9, 1883
Reconstructing the dinosaur genome.

Prondvai, E. 2017. Medullary bone in fossils: Function, evolution and significance in growth curve reconstructions of extinct vertebrates. *Journal of Evolutionary Biology* 30, 440–60.
Where medullary bone occurs, and where it does not occur.

Schweitzer, M., Marshall, M., Carron, K., Bohle, D. S., Busse, S. C., Arnold, E. V., Barnard, D., Horner, J. R., and Starkley, J. R. 1997. Heme compounds in dinosaur trabecular bone. *Proceedings of the National Academy of Sciences, U.S.A.* 94, 6291–96
The first report of dinosaur blood.

Schweitzer, M. H., Wittmeyer, J. L., and Horner, J. R. 2005. Gender-specific reproductive tissue in ratites and *Tyrannosaurus rex. Nature* 308, 1456–60
Report of medullary bone in a giant dinosaur.

Schweitzer, M. H., Wittmeyer, J. L., Horner, J. R., and Toporski, J. K. 2005. Soft-tissue vessels and cellular preservation in *Tyrannosaurus rex. Science* 307, 1952–55

*Shapiro, B. 2015. *How to clone a mammoth: The science of de-extinction.* Princeton University Press, Princeton
A great overview of the whole topic, from Dolly to cloning, and future plans with mammoths.

*Thomas, M., Gilbert, M. T. P., Bandlet, H.-J., Hofreiter, M., and Barnes, I. 2005. Assessing ancient DNA studies. *Trends in Ecology and Evolution* 20, 541–44
A practical overview of the topic.

Wiemann, J., Fabbri, M., Yang, T.-R., Stein, K., Sander, P. M., Norell, M. A., and Briggs, D. E. G. 2018. Fossilization transforms vertebrate hard tissue proteins into N-heterocyclic polymers. *Nature Communications* 9, 4741

Woodward, S. R., Weyand, N. J., and Bunnell, M. 1994. DNA sequence from Cretaceous period bone fragments. *Science* 266, 1229–32
The report of supposed dinosaur DNA.

Benton, M. J., Csiki, Z., Grigorescu, D., Redelstorff, R., Sander, P. M., Stein, K., and Weishampel, D. B. 2010. Dinosaurs and the island rule: The dwarfed dinosaurs from Haţeg Island. *Palaeogeography, Palaeoclimatology, Palaeoecology* 293, 438–54
The dwarf dinosaurs of Transylvania.

*Carpenter, K., Hirsch, K. F., and Horner, J. R. (eds). 1996. *Dinosaur eggs and babies*. Indiana University Press, Bloomington; Cambridge University Press, Cambridge
A series of articles on different examples of dinosaur eggs and babies.

Chapelle, K., and Choiniere, J. N. 2018. A revised cranial description of *Massospondylus carinatus* Owen (Dinosauria: Sauropodomorpha) based on computed tomographic scans and a review of cranial characters for basal Sauropodomorpha. *PeerJ* 6, e4224

*Erickson, G. M. 2005. Assessing dinosaur growth patterns: a microscopic revolution. *Trends in Ecology and Evolution* 20, 677–84
A review of the whole topic.

Erickson, G. M., Curry Rogers, K., and Yerby, S. A. 2001. Dinosaurian growth patterns and rapid avian growth rates. *Nature* 412, 429–33

Erickson, G. M., Makovicky, P. J., Currie, P. J., Norell, M. A., Yerby, S. A., and Brochu, C. A. 2004. Gigantism and comparative life history of *Tyrannosaurus rex*. *Nature* 430, 772–75

Erickson, G. M., Rauhut, O. W. M., Zhou, Z., Turner, A. H., Inouye, B. D., Hu, D., and Norell, M. A. 2009. Was dinosaurian physiology inherited by birds? Reconciling slow growth in *Archaeopteryx*. *PLoS ONE* 4, e7390

Norell, M. A., Clark, J. M., Chiappe, L. M., and Dashzeveg, D. 1995. A nesting dinosaur. *Nature* 378, 774–76
Evidence that *Oviraptor* was the mummy, and that she incubated her eggs.

Reisz, R. R., Scott, D., Sues, H.-D., Evans, D. C., and Raath, M. A. 2005. Embryos of an Early Jurassic prosauropod dinosaur and their evolutionary significance. *Science* 309, 761–64
The *Massospondylus* embryos.

Sander, P. M., Christian, A., Clauss, M., Fechner, R., Gee, C. T., Griebeler, E.-M., Gunga, H.-C., Hummel, J., Mallison, H., Perry, S. F., Preuschoft, H., Rauhut, O. W. M., Remes, K., Tutken, T., Wings, O., and Witzel, U. 2010. Biology of the sauropod dinosaurs: the evolution of gigantism. *Biological Reviews* 86, 117–55

Zhao, Q., Benton, M. J., Sullivan, C., Sander, P. M., and Xu, X. 2013. Histology and postural change during the growth of the ceratopsian dinosaur *Psittacosaurus lujiatunensis*. *Nature Communications* 4, 2079

Chapter 7
How Did Dinosaurs Eat? (pp. 186–214)

*Barrett, P. M., and Rayfield, E. J. 2006. Ecological and evolutionary implications of dinosaur feeding behaviour. *Trends in Ecology and Evolution* 21, 217–24
A review of how palaeontologists determine dinosaur feeding behaviour.

Bates, K. T., and Falkingham, P. L. 2012. Estimating maximum bite performance in *Tyrannosaurus rex* using multi-body dynamics. *Biology Letters* 8, 660–64

Button, D. J., Rayfield, E. J., and Barrett, P. M. 2014. Cranial biomechanics underpins high sauropod diversity in resource-poor environments. *Proceedings of the Royal Society B* 281, 20142114
Resource partitioning among the Morrison sauropods.

Chin, K., and Gill, B. D. 1996. Dinosaurs, dung beetles, and conifers: Participants in a Cretaceous food web. *Palaios* 11, 280–85

Chin, K., Tokaryk, T. T., Erickson, G. M., and Calk, L. 1998. A king-sized theropod coprolite. *Nature* 393, 680–82

Erickson, G. M., Krick, B. A., Hamilton, M., Bourne, G. R., Norell, M. A., Lilleodden, E., et al. 2012. Complex dental structure and wear biomechanics in hadrosaurid dinosaurs. *Science* 338, 98–101

Gill, P. G., Purnell, M. A., Crumpton, N., Brown, K. R., Gostling, N. J., Stampanoni, M., and Rayfield, E. J. 2014. Dietary specializations and diversity in feeding ecology of the earliest stem mammals. *Nature* 591, 303–5

Godoy, P. L., Montefeltro, F. C., Norell, M. A., and Langer, M. C. 2014. An additional baurusuchid from the Cretaceous of Brazil with evidence of interspecific predation among Crocodyliformes. *PLoS ONE* 9(5), e97138
The crocodilian-dominated foodweb of the Adamantina Formation.

Mitchell, J. S., Roopnarine, P. D., and Angielczyk, K. D. 2012. Late Cretaceous restructuring of terrestrial communities facilitated the end-Cretaceous mass extinction in North America. *Proceedings of the National Academy of Sciences, U.S.A.* 109, 18857–61

Rayfield, E. J. 2004. Cranial mechanics and feeding in *Tyrannosaurus rex. Proceedings of the Royal Society B* 271, 1451–59

Rayfield, E. J. 2005. Aspects of comparative cranial mechanics in the theropod dinosaurs *Coelophysis, Allosaurus* and *Tyrannosaurus. Zoological Journal of the Linnean Society* 144, 309–16

*Rayfield, E. J. 2007. Finite element analysis and understanding the biomechanics and evolution of living and fossil organisms. *Annual Review of Earth and Planetary Sciences* 35, 541–76
A review of the FEA method as applied to dinosaurs and other fossil animals.

Rayfield, E. J., Milner, A. C., Xuan, V. B., and Young, P. G. 2007. Functional morphology of spinosaur 'crocodile-mimic' dinosaurs. *Journal of Vertebrate Paleontology* 27, 892–901

Rayfield, E. J., Norman, D. B., Horner, C. C., Horner, J. R., May Smith, P., et al. 2001. Cranial design and function in a large theropod dinosaur. *Nature* 409, 1033–37

Chapter 8
How Did They Move and Run?

(pp. 215–53)

Alexander, R. McN. 1976. Estimates of speeds of dinosaurs. *Nature* 261, 129–30

*Alexander, R. McN. 1989. *Dynamics of dinosaurs and other extinct giants.* Columbia University Press, New York
Still an excellent introduction.

*Alexander, R. McN. 2006. Dinosaur biomechanics. *Proceedings of the Royal Society B* 273, 1849–55
The master speaks.

Bishop, P. J., Graham, D. F., Lamas, L. P., Hutchinson, J. R., Rubenson, J., Hancock, J. A., Wilson, R. S., Hocknull, S. A., Barrett, R. S., Lloyd, D. G., et al. 2018. The influence of speed and size on avian terrestrial locomotor biomechanics: Predicting locomotion in extinct theropod dinosaurs. *PLoS ONE* 13, 0192172

Coombs, W. P., Jr. 1980. Swimming ability of carnivorous dinosaurs. *Science* 207, 1198–1200

Falkingham, P. L., and Gatesy, S. M. 2014. The birth of a dinosaur footprint: Subsurface 3D motion reconstruction and discrete element simulation reveal track ontogeny. *Proceedings of the National Academy of Sciences, U.S.A.* 111, 18279–84

Galton, P. M. 1970. The posture of hadrosaurian dinosaurs. *Journal of Paleontology* 44, 464–73

Gatesy, S. M., Middleton, K. M., Jenkins, F. A., and Shubin, N. H. 1999. Three-dimensional preservation of foot movements in Triassic theropod dinosaurs. *Nature* 399, 141–44

*Gillette, G. G., and Lockley, M. G. (eds). 1989. *Dinosaur tracks and traces*. Indiana University Press, Bloomington; Cambridge University Press, Cambridge
An overview and many case studies.

*Haines, T. 1999. *Walking with dinosaurs: A natural history*. BBC Books, London; DK, New York
The producer of the series talks about animation methods and ensuring accuracy.

Heers, A. M., and Dial, K. P. 2015. Wings versus legs in the avian bauplan: Development and evolution of alternative locomotor strategies. *Evolution* 69, 305–20

Henderson, D. M. 2006. Burly gaits: Centers of mass, stability, and the trackways of sauropod dinosaurs. *Journal of Vertebrate Paleontology* 26, 907–21

Hutchinson, J. R., and Garcia, M. 2002. *Tyrannosaurus* was not a fast runner. *Nature* 415, 1018–21

Hutchinson, J. R., and Gatesy, S. M. 2006. Dinosaur locomotion: Beyond the bones. *Nature* 440, 292–94

Kubo, T., and Benton, M. J. 2009. Tetrapod postural shift estimated from Permian and Triassic trackways. *Palaeontology* 52, 1029–37
The switch from sprawlers to upright walkers across the Permian–Triassic mass extinction.

Lockley, M. G., Houck, K., and Prince, N. K. 1986. North America's largest dinosaur tracksite: Implications for Morrison Formation paleoecology. *Geological Society of America, Bulletin* 97, 1163–76

Mickelson, D., King, M., Getty, P., and Mickelson, K. 2006. Subaqueous tetrapod swim tracks from the middle Jurassic Bighorn Canyon National Recreation Area (BCNRA), Wyoming, USA. *New Mexico Museum of Natural History and Science Bulletin* 34
Summary only: the full paper has not been published.

*Ostrom, J. H. 1979. Bird flight: How did it begin? *American Scientist* 67, 46–56
The classic 'ground up' view for the origin of bird flight.

Palmer, C. 2014. The aerodynamics of gliding flight and its application to the arboreal flight of the Chinese feathered dinosaur *Microraptor*. *Biological Journal of the Linnean Society* 113, 828–35

*Xu, X., Zhou, Z., Dudley, R., et al. 2014. An integrative approach to understanding bird origins. *Science* 346, 1253293
A current review of bird origins and the 'trees down' model for the origin of flight.

Chapter 9
Mass Extinction

*Alvarez, L. W., Alvarez, W., Asaro, F., and Michel, H. V. 1980. Extraterrestrial cause for the Cretaceous–Tertiary extinction. *Science* 208, 1095–1108
The original proposal of impact.

*Alvarez, W. 2008. *T. rex and the crater of doom*, 2nd edition. Princeton University Press, Princeton
Arguably the best title ever for a popular science book – Walter Alvarez tells the whole story.

Benton, M. J. 1990. Scientific methodologies in collision: The history of the study of the extinction of the dinosaurs. *Evolutionary Biology* 24, 371–424
The 100 top reasons for dinosaur extinction.

Field, D. J., Bercovici, A., Berv, J. S., Dunn, R. E., Fastovsky, D. E., Lyson, T. R., Vajda, V., and Gauthier, J. A. 2018. Early evolution of modern birds structured by global forest collapse at the end-Cretaceous mass extinction. *Current Biology* 28, 1825–31

Hildebrand, A. R., Penfield, G. T., Kring, D. A., Pilkington, M., Camargo, A., Jacobsen, S. B., and Boyton, W. V. 1991. Chicxulub crater – a possible Cretaceous/Tertiary boundary impact crater on the Yucatán Peninsula, Mexico. *Geology* 19, 867–71

Lyell, C. 1830–33. *Principles of geology, being an attempt to explain the former changes of the Earth's surface, by reference to causes now in operation*, 3 vols. John Murray, London
The classic expression of uniformitarianism.

MacLeod, K. G., Quinton, P. C., Sepúlveda, J., and Negra, M. H. 2018. Postimpact earliest Paleogene warming shown by fish debris oxygen isotopes (El Kef, Tunisia). *Science* 24, eaap8525

Maurrasse, F. J.-M. R., and Sen, G. 1991. Impacts, tsunamis, and the Haitian Cretaceous-Tertiary boundary layer. *Science* 252, 1690–93

Morgan, J., Warner, M., Brittan, J., Buffler, R., Camargo, A., Christeson, G., Dentons, P., Hildebrand, A., Hobbs, R., MacIntyre, H., Mackenzie, G., Maguires, P., Marin, L., Nakamura, Y., Pilkington, M., Sharpton, V., and Snyders, D. 1997. Size and morphology of the Chicxulub impact crater. *Nature* 390, 472–76

Raup, D. M., and Sepkoski, J. J., Jr. 1984. Periodicity of extinctions in the geologic past. *Proceedings of the National Academy of Sciences, U.S.A.* 81, 801–05

Sakamoto, M., Benton, M. J., and Venditti, C. 2016. Dinosaurs in decline tens of millions of years before their final extinction. *Proceedings of the National Academy of Sciences, U.S.A.* 113, 5036–40

Slater, G. J. 201. Phylogenetic evidence for a shift in the mode of mammalian body size evolution at the Cretaceous–Palaeogene boundary. *Methods in Ecology and Evolution* 4, 734–44

Wolfe, J. A. 1991. Palaeobotanical evidence for a June 'impact' at the Cretaceous/Tertiary boundary. *Nature* 352, 420–23

Chapter 10
Afterword (pp. 286–89)

Maddox, J. 1998. *What remains to be discovered.* The Free Press, New York; Macmillan, London

Oreskes, N., and Conway, E. M. 2010. *Merchants of doubt.* Bloomsbury, London and New York
The misuse of science by scientists to make political points, especially by the former pro-smoking lobby, and currently by the climate-change deniers.

General books about dinosaurs

Benton, M. J. 2015. *Vertebrate palaeontology*, 4th edition. Wiley, New York and Oxford
The standard textbook on the subject, putting dinosaurs in context of all other vertebrates.

Brett-Surman, M., Holtz, T., Jr., and Farlow, J. O. (eds). 2012. *The complete dinosaur*, 2nd edition. Indiana University Press, Bloomington
Over 1000 pages of articles on every subject to do with dinosaurian discovery, evolution, and biology.

Brusatte, S. L. 2012. *Dinosaur paleobiology*. Wiley, New York and Oxford
An excellent serious introduction to dinosaurs.

Brusatte, S. L. 2018. *The rise and fall of the dinosaurs: the untold story of a lost world*. Macmillan, New York and London
The best recent narrative of the life of a dinosaur hunter.

Fastovsky, D. E., and Weishampel, D. B. 2016. *Dinosaurs: a concise natural history*, 3rd edition. Cambridge University Press, Cambridge
The best standard textbook on dinosaurs.

Klein, N., Remes, K., Gee, C., and Sander, P. M. (eds). 2011. *Biology of the sauropod dinosaurs: understanding the life of giants*. Indiana University Press, Bloomington
The most thorough consideration of the biology of the largest dinosaurs.

Naish, D., and Barrett, P. M. 2018. *Dinosaurs: how they lived and evolved*. Natural History Museum, London
Dinosaurs in full colour – history, diversity, and palaeobiology.

Norell, M. A. 2019. *The world of dinosaurs: the ultimate illustrated reference*. University of Chicago Press, Chicago
Collecting and studying dinosaurs, and with insights from the great work at the AMNH.

Weishampel, D. B., Dodson, P., and Osmolska, H. (eds). 2004. *Dinosauria*, 2nd edition. University of California Press, Berkeley
A few years old now, but the most comprehensive account of the diversity of dinosaurs.

White, S. 2012. *Dinosaur art: the world's greatest paleoart*. Titan Books, New York
An introduction to the amazing efforts of artists to reconstruct the dinosaurs.

Illustration Credits

Index